Apple Pro Training Series

Final Cut Pro 6 for News and Sports Quick-Reference Guide

Joe Torelli

Apple Pro Training Series: Final Cut Pro 6 for News and Sports Quick-Reference Guide
Joe Torelli
Copyright © 2007 Joe Torelli

Published by Peachpit Press. For information on Peachpit Press books, contact:
Peachpit Press
1249 Eighth Street
Berkeley, CA 94710
(510) 524-2178
(800) 283-9444
Fax: (510) 524-2221
http://www.peachpit.com
To report errors, please send a note to errata@peachpit.com
Peachpit Press is a division of Pearson Education

Technical Reviewers: Bob Craigue, Bob Crowley, Nathan Haggard, Ted Phillips

Notice of Rights
All rights reserved. No part of this book may be reproduced or transmitted in any form by any means, electronic, mechanical, photocopying, recording, or otherwise, without the prior written permission of the publisher. For information on getting permission for reprints and excerpts, contact permissions@peachpit.com.

KCBS/KCAL images courtesy of KCBS-TV/KCAL9, Los Angeles. Used by permission. CBS, CBS2 KCBS/KCAL or the CBS Eye Design are registered trademarks of CBS Broadcasting, Inc. All rights reserved.

Notice of Liability
The information in this book is distributed on an "As Is" basis, without warranty. While every precaution has been taken in the preparation of the book, neither the author nor Peachpit Press shall have any liability to any person or entity with respect to any loss or damage caused or alleged to be caused directly or indirectly by the instructions contained in this book or by the computer software and hardware products described in it.

Trademarks
Many of the designations used by manufacturers and sellers to distinguish their products are claimed as trademarks. Where those designations appear in this book, and Peachpit Press was aware of the trademark claim, the designations appear as requested by the owner of the trademark. All other product names and services identified throughout the book are used in an editorial fashion only and for the benefit of such companies with no intention of infringement of the trademark. No such use, or the use of any trade name, is intended to convey endorsement or other affiliation with this book.

ISBN 13: 9780321514233

ISBN 10: 0-321-51423-8

9 8 7 6 5 4 3 2 1

Printed and bound in the United States of America

Table of Contents

Chapter 1	A Quick Tour of Final Cut Pro............ 1	
	Interface Overview 2	
	The Browser......................................3	
	The Viewer.......................................6	
	The Canvas 7	
	The Timeline 8	
	Tool Palette 9	
	Audio Meters..................................10	
Chapter 2	Initial Setup 11	
	Choosing a Format............................14	
Chapter 3	Log and Transfer from Tapeless Media ... 19	
	Overview of Tapeless Acquisition Devices...........19	
	Panasonic P2 36	
	Ikegami Editcam...............................43	
	Focus Enhancements FireStore 45	
Chapter 4	Log and Capture from Videotape 47	
	Logging Videotape Prior to Capture................48	
	Preparing to Capture from Videotape 49	
	Capturing from Videotape......................54	
Chapter 5	Viewing Media with Final Cut Server 57	
	The Producer or Reporter's Workflow 58	
	The Editor's Workflow 62	
Chapter 6	Editing Simple Voiceovers............... 65	
	Storyboard Edit................................66	
	Rearrange in the Timeline 69	
	Using the Viewer with Clips from Videotape........72	
	Using the Viewer with Tapeless Media 73	

iii

iv   Contents

Topping and Tailing 73
Simple Audio Adjustments. 79
Audio Cross Fades. 85
Adding Dissolves. 88
Fixing Material 90

Chapter 7 Fast Package Editing 91
Organizing for the Edit 91
Laying Out the Story 92
Adding B-roll. 103
Backtiming. 109
Padding the End of a Sequence 110
Opening for NATSOT. 112
Natural Sound "Pops". 115
Video Effects and Rendering 119
Finishing 119

Chapter 8 Basic Fixes. 121
Motion Effects. 121
Fixing White Balance 127
Blurring a Subject 128
Highlighting a Subject 133
Resize 134
Adding Other Effects 136
Image Stabilization—SmoothCam 137
Audio Equalization 138

Chapter 9 Delivering the Story. 141
Preparing for Air. 141
Live to Air 147
Playback to Tape 148
Remote Story Delivery 148

Chapter 10 Customization Overview 173
Creating Keyboard Shortcuts. 173
Configuring Button Bars 176
Customizing the Window Layout 179

Index 182

1
A Quick Tour of Final Cut Pro

Preparing news and sports video for broadcast and Web delivery requires getting material into your system, edited, and out again as fast as possible. Deadlines are within minutes, not days or weeks. This book focuses only on the keystrokes and menu commands essential to delivering your stories as fast as possible using Final Cut Pro. Whether you are a field journalist working on a MacBook Pro and sending material back via FTP or you're crashing a story under a tight deadline using networked storage, the methods in this book will help you get your stories done quickly and with high quality.

This book provides overviews of the Final Cut Pro interface, buttons, keys, and menus, but the centerpieces are Chapter 6, "Editing Simple Voiceovers," and Chapter 7, "Fast Package Editing." These chapters give you the information you'll need to quickly edit these two story types. To help you deal with the practical requirement of going tapeless, you'll learn how to work not just with videotape but also P2, XDCAM HD, and other tapeless technologies. To better prepare you for working from any location, you'll learn how to get your stories delivered from both the comfort of your facility and from remote locations using methods such as sat phone. (If you want a general guide to the full functionality of Final Cut Pro, refer to *Apple Pro Training Series: Final Cut Pro* by Peachpit Press.)

To get you started, this chapter presents an overview of what you see on the Final Cut Pro screen.

Interface Overview

Familiarize yourself with the basic look and feel of the Final Cut Pro interface. Getting comfortable with the windows and buttons will help you get your bearings and gain speed in the editing process.

The Final Cut Pro user interface (UI) is divided into several windows. Each can be positioned where you want it, and most of the windows can be resized.

The Browser contains the projects, clips, and sequences you will combine and edit to create your final segment.

The Viewer is used to preview individual clips in the Browser or detail of an edited sequence; to set In and Out points, markers, and keyframes; and to create and adjust motion effects.

The Canvas displays the frame that's selected or active in the Timeline and allows you to fine-tune your edits and effects placement. Changes made in the Canvas are immediately applied in the Timeline, and vice-versa; closing either window also closes the other.

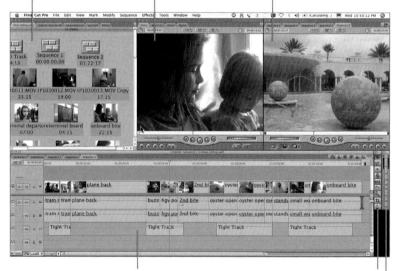

The Timeline is used for sequencing video and audio clips and for adding transitions, titles, and effects.

The Tool palette lets you change the pointer into one of many tools you can use interactively to edit and adjust clips in the Timeline.

The Audio Level meters provide a real-time visual representation of the audio portion of a clip or a sequence as it plays.

Standard Mac OS X interface conventions apply in Final Cut Pro:

- To move a window, click its title bar and drag the window to a new position.
- To resize a window, drag the window's lower-right corner.

The Browser

The Browser is the primary location for organizing your *clips*, or individual video and audio segments, and *sequences*, or groups of edited clips, ordered the way you want them to appear. You can set up the Browser in many ways, according to your needs, and you can easily toggle the Browser between List and Frame views (similar to the List and Icon views in the Mac OS X Finder).

Each tab bears the name of an open (active) project.

To select and use a clip stored in the Browser, drag it to the Timeline. To copy a clip or sequence to another active project, drag it onto the other project's tab in the Browser.

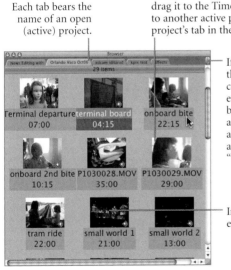

If you drag command buttons onto the button bar (also known as the coffee bean), it separates to become end brackets for your custom button bar. For more information about the coffee bean, which also appears in the Viewer, Canvas, and Timeline, see Chapter 10, "Customization Overview."

In Frame view, thumbnails display each clip's name and duration.

Final Cut Pro organizes video elements into *bins*, which are comparable to Mac OS X folders (and which have icons that look like folders). Use bins to organize material by story, by topic, or however else you prefer to categorize your content.

4 A Quick Tour of Final Cut Pro

The Browser displays one open project, named "News editing."

The B-Roll bin is opened as a separate window, and its contents are displayed.

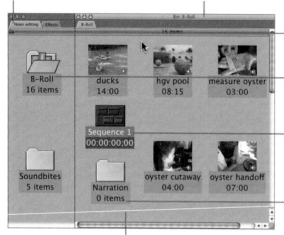

This icon shows that this is the active Logging bin.

The icon for the bin called B-Roll indicates that the bin contains 16 items and is open.

The Sequence 1 icon's zero-duration label means it is a new sequence that contains no clips or edits.

The Narration bin's icon indicates that the bin is closed and empty (it contains zero items).

The open window for the B-Roll bin displays the clips and sequences stored in the bin. This window, in Icon view, shows 7 of the bin's 16 items; scrolling would reveal the rest.

Effects Tab

The Browser includes an Effects tab. Click this tab to view your audio and video effects bins and their contents. If you inadvertently close the Effects tab in the Browser, you can access it again by choosing it from the Window menu, or by pressing Command-5.

Selecting the Effects tab displays the available video and audio effects folders and their contents. Here, the folders and effects are shown in List view.

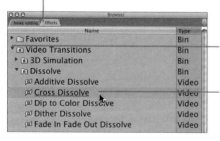

As with the Mac OS X Finder, you can click the disclosure triangle next to a bin's name to view the bin's contents.

The underlined effect is the default effect, which for a video transition can be applied by pressing Command-T.

Browser Options

Final Cut Pro, much like the Finder, lets you view bin contents in multiple ways.

You can change many of the Browser options via contextual shortcut menus. Control-click (or right-click) a bin window to view the options.

The Set Logging Bin command lets you assign the default target location for your clips and sequences. When you capture material from tape or via transfer, Final Cut Pro places it in the Logging bin.

You can view the contents of the Browser as small-, medium-, or large-sized icons, or by name in a sortable list view.

When the Browser is in List view, clicking a column header sorts the list by that column's contents. Click a header once to sort in ascending order; click again to reverse the sort to descending order.

You can see timecode, duration, marks, tracks, and other information in the List view by scrolling or by resizing the Browser.

You may want certain columns clustered together to see them more easily. Drag a column header left or right and drop it where you want that column positioned.

You can create new bins and new sequences from the File menu, or you can use these shortcuts:

▶ Press Command-B to create a new bin.

▶ Press Command-N to create a new sequence.

The Viewer

The Viewer is not just what most non-linear editors consider to be the "source" side, as in a tape edit room. It is much more. Of course you can use it to load clips and screen them, select how much of a clip to use by marking In and Out points, but you can also load clips in a sequence in the Viewer to modify them in many different ways. For instance, you can apply a slow-motion effect or resize the image by double-clicking the element in the Timeline and opening the Motion tab in the Viewer.

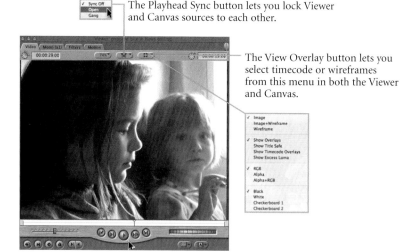

This readout indicates the duration of the portion of your clip that you've marked with In and Out points for capture, sequencing, or other manipulation. If you haven't marked anything, it displays the full duration of the current active clip. If you have a Mark-In it displays the duration from the mark to the last frame. If you have a Mark-Out, it displays the duration from the first frame to your mark.

This readout shows the current timecode position.

Use the shuttle control to speed through your clip preview.

The playhead shows where you are in the clip or sequence.

You can use these tools to navigate through the clip to find suitable In and Out points, or use keyboard shortcuts to do so.

Use the jog control for frame-by-frame navigation within a clip.

Mark In (I) and
Mark Out (O)
Add Marker (M)
Add Keyframe (Option-K)
Mark Clip (X)
Show Match Frame (F)

You can easily access generator files using the Generator pop-up menu.

Use the Recent Clip menu for quick access to recent clips you have viewed. The number of clips can be set using the User Preferences General setting "List recent clips".

The Canvas

This is the visual representation of what is in the Timeline. You may drag the playhead in either the Timeline or the area directly under the Canvas to preview your sequence. When you play the Timeline, you will see the sequence play in the Canvas.

The Canvas looks similar to the Viewer, but it always displays the content at the Timeline playhead. It also contains several editing function buttons that are not found in the Viewer.

The Playhead Sync button works just as it does in the Viewer.

The View overlay button works the same as in the Viewer.

Click the Insert Edit button to perform an insert, or splice edit, splitting the clip in the Canvas at the selected point and inserting material selected in the Viewer. The inserted content will push the remainder of the target sequence forward, thereby lengthening the sequence.

Click the Replace Edit button after positioning the playhead in the Timeline or Canvas and selecting a new sync point in the Viewer. The shot in the Viewer will automatically replace the remainder of the marked Timeline or Canvas sequence.

Click the Overwrite Edit button to cover over the portion of the marked area of the Canvas with whatever clip is selected in the Viewer. If you apply this edit to an existing marked shot, that shot's duration will be preserved. If the marked area in the source clip is shorter than the marked area in the Timeline or Canvas, an "insufficient duration" warning appears. You can then re-mark the source clip or perform a fit-to-fill edit.

The Timeline

The Timeline is the primary graphical representation of the edited sequence. It is laid out left to right, with the first shot beginning on the left and progressing to the last shot on the right.

Tool Palette

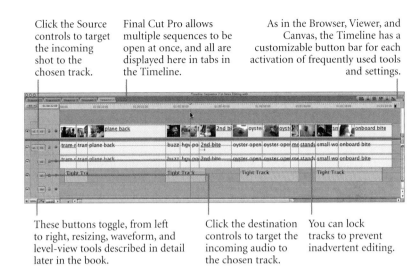

Click the Source controls to target the incoming shot to the chosen track.

Final Cut Pro allows multiple sequences to be open at once, and all are displayed here in tabs in the Timeline.

As in the Browser, Viewer, and Canvas, the Timeline has a customizable button bar for each activation of frequently used tools and settings.

These buttons toggle, from left to right, resizing, waveform, and level-view tools described in detail later in the book.

Click the destination controls to target the incoming audio to the chosen track.

You can lock tracks to prevent inadvertent editing.

Tool Palette

This palette lets you quickly choose from among the most-used edit functions. Holding down a button with an arrow at the top right reveals more buttons related to that particular function. For news and sports projects, some will rarely be used. As you begin editing more often with Final Cut Pro, you will find yourself relying less on the mouse and more on the keyboard. The keys that typically get the most use are the following:

A	Selection tool
J	Press once for play speed in reverse. Press again, 2x; again, 4x
K	With the J key, jogs in reverse at 1/3 speed. With L, jogs forward at 1/3 speed
L	Press once for play speed forward. Press again, 2x; again, 4x
I	Marks an in point
O	Marks an out point
X	Places marks at the first and last frame of an edit
N	Turns snapping on or off, even while dragging
Q	Toggles between the Viewer and the Canvas/Timeline
R	Roll edit – adjusts transition points earlier or later
RR	Ripple edit – shortens or lengthens edits at the transition point

10 A Quick Tour of Final Cut Pro

U Cycles outgoing/centered/incoming transition point
E Extends to the playhead if a highlighted transition is located before the playhead. Conversely, this key "backtimes" the next highlighted transition if the playhead is positioned before that transition.
S Slip edit—adjusts the first and last frame within an edit and keeps the shot's duration the same, as well as its position in the Timeline
SS Slide edit—moves a selected clip earlier or later in the sequence
B Blade—breaks a shot into two pieces
P Pen tool—clicking on a track in the Timeline or Viewer adds a keyframe to that track

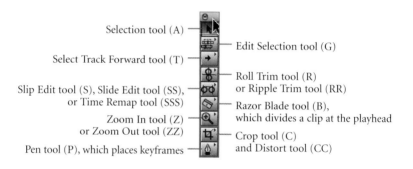

Selection tool (A)
Select Track Forward tool (T)
Slip Edit tool (S), Slide Edit tool (SS), or Time Remap tool (SSS)
Zoom In tool (Z) or Zoom Out tool (ZZ)
Pen tool (P), which places keyframes
Edit Selection tool (G)
Roll Trim tool (R) or Ripple Trim tool (RR)
Razor Blade tool (B), which divides a clip at the playhead
Crop tool (C) and Distort tool (CC)

Audio Meters

These meters display audio levels in real time as content plays in the Timeline, Canvas, or Viewer.

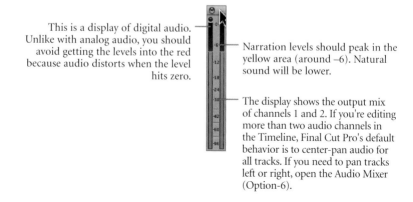

This is a display of digital audio. Unlike with analog audio, you should avoid getting the levels into the red because audio distorts when the level hits zero.

Narration levels should peak in the yellow area (around –6). Natural sound will be lower.

The display shows the output mix of channels 1 and 2. If you're editing more than two audio channels in the Timeline, Final Cut Pro's default behavior is to center-pan audio for all tracks. If you need to pan tracks left or right, open the Audio Mixer (Option-6).

2
Initial Setup

To take full advantage of Final Cut Pro, you need to make certain preparations before you begin your work. If you are the only person using your Mac, you may not need to change or confirm any settings. If, however, you are sharing a system with others, you need to ensure that the compression and format settings are configured properly for your edit. Remember: You work under tight deadlines, and the last thing you need is to be unable to deliver your story once you finish editing.

The Open Format Timeline introduced in Final Cut Pro 6 greatly simplifies the process of editing multiple formats, compression types, and frame rates in the same sequence. Also, you can mix standard definition and high definition material in the same sequence and play it. With Open Format Timeline, you literally drag or edit the first shot into the Timeline, and the Timeline automatically adapts to its format, compression, and frame-rate settings. Each additional clip you add to the Timeline, regardless of its native formatting, automatically adopts those "master settings." Support for a mixture of clips with different formats and compression settings in the same Timeline gives you the ability to play each edit without having to conform or render those shots during the creative process.

The Open Format Timeline is always enabled. Depending on your playout method (from the Timeline or exporting to a playout or Web server), if you are mixing formats, you may want to start a new sequence with a shot from the dominant format for your story. This will adapt the Timeline to conform to that setting and will reduce render times that may be necessary for any shots that aren't the same format. If you are going to export to a playout server, you will most likely need to match the server's format and compression. To do this, open Easy Setup, choose Final Cut Pro > Easy Setup or press Control-Q.

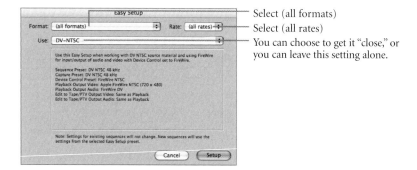

Select (all formats)
Select (all rates)
You can choose to get it "close," or you can leave this setting alone.

Once your server's target format and compression are selected, it's a good idea to edit the first shot in the Timeline with a shot that matches this setting. This prevents the Open Timeline from changing the sequence to a format or compression that doesn't match the playout server, which could delay getting your story to air.

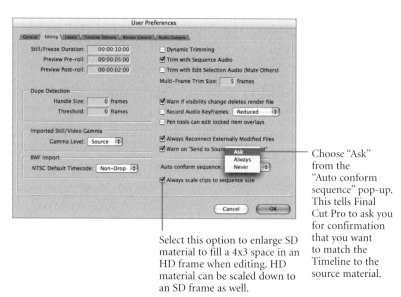

Choose "Ask" from the "Auto conform sequence" pop-up. This tells Final Cut Pro to ask you for confirmation that you want to match the Timeline to the source material.

Select this option to enlarge SD material to fill a 4x3 space in an HD frame when editing. HD material can be scaled down to an SD frame as well.

Initial Setup   13

When the "Auto conform sequence" setting is set to Ask, this dialog will appear when a shot that doesn't match the target compression and frame rate is placed in the Timeline. Click Yes to make the Timeline's settings automatically match those of the first clip being edited.

If you want to retain the Timeline settings (as mentioned earlier with a playout server), click No. This allows you to edit a shot that is different in format and compression from what you chose in Easy Setup.

This is true for common formats as well as obscure formats such as QuickTime movies recorded on still cameras using Photo-JPEG compression. To check the source material type after editing the first shot, open the Sequence Settings (press Command-0).

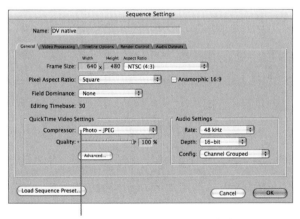

Despite the use of non-standard Photo-JPEG compression at 640x480, the Open Format Timeline auto-detects the source material and matches the Timeline to it, avoiding lengthy trial-and-error attempts to match it manually.

If working as a standalone editing environment where you are not connected to shared storage such as Apple Xsan or using playout servers from manufacturers such as Omneon, Grass Valley, or Leitch, you need only ensure that the capture presets and the Timeline match to ensure good performance.

If you are connected to shared storage or a playout server, decisions about compression and frame rate have already been made for you by the engineering department. The data being edited in these server-based environments is just that, data. The data is then converted to video at the playout server, instead of at the Final Cut Pro editor.

Choosing a Format

Final Cut Pro works equally well for both high definition (HD) and standard definition (SD) formats. There are many flavors of each, so having some understanding of the varieties is helpful.

High Definition

There are several popular compression methods for HD video, including DVCPRO HD (sometimes referred to as DVCPRO 100), HDV, HDCAM, and now Apple's ProRes 422.

The DVCPRO HD format captures video at a rate of 100 megabits per second (Mbit/s) and compresses every frame individually.

HDV is a prosumer (or professional and consumer) format that has been accepted by several manufacturers. It also has a professional cousin that is used as a compression format for Sony's XDCAM HD professional disk acquisition system. HDV and XDCAM HD can capture at 25 Mbit/s, but they employ a compression scheme called Long Group of Pictures (Long GoP) MPEG, which selectively discards some video frames. XDCAM HD also can record at 18 Mbit/s and, at its highest-quality setting, 35 Mbit/s. HDV can record at only 25 Mbit/s.

HDCAM from Sony is an origination format/codec that mildly compresses the image and captures every frame, unlike XDCAM HD. Editing in FCP requires a video capture card capable of capturing an HD-SDI signal.

With Final Cut Pro 6, Apple has released a high-quality mild compression production codec, ProRes 422. You can transcode to ProRes 422 or capture as video using the HD-SDI input of a capture card inside the Mac, or a FireWire attached device.

These HD standards support multiple *delivery formats*, or combinations of output frame size and frame rate. The most important of these for broadcast use in North America are 1080i60 and 720p60.

In the U.S., the CBS and NBC networks generally broadcast at 1080ip60, which denotes a vertical frame size of 1,080 *interlaced* lines that are rendered at a rate of 60 fields per second. Interlacing is an output method that draws each frame on the screen in two passes, or *fields,* by rendering the even-numbered rows and then the odd-numbered rows. (Because each field consists of half the content of a frame, a render rate of 60 fields per second is equivalent to a rate of 30 frames per second.)

The ABC and Fox networks typically broadcast at 720p60, which denotes a frame height of 720 vertical lines rendered *progressively* at a rate of 60 frames per second. Progressive rendering draws each line of a frame in succession, from top to bottom. Most computer displays are progressive.

If you work for a newspaper delivering HD material to a Web site, you most likely will be working in 720p30 format, which is well-suited to the way LCD and plasma screens and projectors handle progressive images. This is also a good format for podcast delivery for use with Apple TV on home theater projectors and large screens.

If you work for a network or station, the necessary capture standards and delivery formats likely will be constant and will be set up for you in Final Cut Pro by your technical team. If you are working in a standalone environment or if you supply video to multiple networks or stations, identify the delivery format that your customer needs and then work backward to determine the format in which you should acquire.

For instance, if you are delivering to a CBS station in HD, you should confirm that you should be working in 1080i60. Then set your camera to 1080i60, capture everything at 1080i60, and continue the post-production process accordingly in Final Cut Pro. The only variables you need to consider are the type and amount of compression.

Be sure to confirm the acquisition and delivery formats. Some station groups may include stations that span the various HD formats. A station group may have a unified standard for all its stations, so that, for instance, an ABC affiliated station may be acquiring at 1080i instead of the expected 720p.

If you are always going to be working in the same format, it is a good idea to set the Easy Setup to the target compression you need. As mentioned earlier, you can reject the Open Timeline auto-detect feature by clicking No when dropping a nonmatching shot into a preset Timeline.

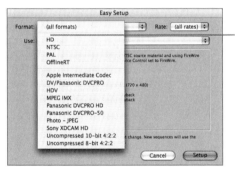

Clicking the Format pop-up displays the groups of compression and formats. Any choice other than "(all formats)" reduces the available options for your Timeline preset. If you work only in HD, select it. This will reduce the choices, excluding any SD-only formats and compression choices.

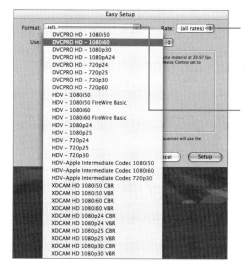

By selecting a specific frame rate from the Rate menu, you further reduce the choices available and narrow your target setting.

The Use menu lets you choose from among the remaining options that comply with the other choices you have made.

In most cases, once your system is set up properly for a project, you will not need to change it. However, it is always a good idea to confirm your settings so as to not have last minute surprises.

Standard Definition

Sony and Panasonic have dominated the broadcast-videotape SD market for decades, going back to the mid-1970s and Sony's U-Matic, which recorded onto ¾-inch tape and introduced electronic editing. Sony released many flavors of its Betacam over the years and introduced compression with its Digital Betacam and SX. Sony's current format, IMX, can be recorded onto videotape or onto XDCAM disks with 30, 40, and 50 Mbit/s compression rates.

Panasonic's DVCPRO SD format has a capture compression rate of 25 Mbit/s and also supports higher-quality capture at 50 Mbit/s. The DVCPRO standard is optimized for delivery to videotape or to P2 static RAM cards.

Sony has a format similar to DVCPRO, called DVCAM, which captures at only 25 Mbit/s. DVCAM can be recorded on videotape or on XDCAM disks.

If you select NTSC from the Format menu, only appropriate frame rates and compression schemes appear in the Use pop-up. These are all the choices available in SD. Other choices include PAL and OfflineRT.

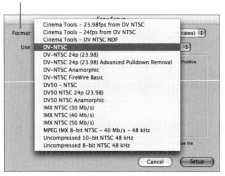

Consumer and prosumer devices shoot DV25, which is also similar in architecture to DVCPRO and DVCAM at 25Mbit/s.

Videotape content is captured in real time in Final Cut Pro through either the FireWire port on a camera or deck, or by using an internal video capture card, such as those from AJA and BlackMagic Design. For example, capturing a 17-minute tape takes 17 minutes. Higher-speed file transfers are available with the disk- and RAM-based acquisition methods, discussed in Chapter 3.

3
Log and Transfer from Tapeless Media

While videotape is still prevalent in most news and sports operations, tapeless capture is gaining momentum for its speed and flexibility. Even if you use tape exclusively today, a basic understanding of tapeless media will ease the inevitable transition and prepare you for the day you encounter material in a tapeless format from a sister station or organization.

Overview of Tapeless Acquisition Devices

Several devices can capture directly onto tapeless media. These include integrated cameras with disk- or RAM-based recording like the Sony XDCAM, Panasonic P2, and the Ikegami Editcam. All three of these product lines include cameras and decks that support standard definition (SD) video, high definition (HD), or both.

These tapeless devices have several things in common, one of which is that they do not capture natively in the QuickTime format required by Final Cut Pro. Before FCP can work with the captured files, you must "rewrap" the clips, a process by which the files stay untouched but the way FCP looks at them changes. Rewrapping occurs transparently as FCP (or a third-party plug-in or utility) imports your clips from the tapeless storage device. For P2 and Editcam, this process is very fast, as the throughput of these devices is high. For XDCAM and its high definition sibling, XDCAM HD, the process may occur just two or three times faster than standard playback. Rewrapping leaves the video and audio "essence" untouched and effectively indexes the file so it can be used as a QuickTime file once it has been imported.

Another family of devices, Focus Enhancements' Direct To Edit FireStore recorders, lets you capture direct to disk from tape-based or other disk-based cameras. Connecting a FireStore recorder to a videotape camera's FireWire port lets you shoot to tape and to disk at the same time. FireStore has native QuickTime support, so the clips it captures are ready to edit as soon as you plug the FireStore into a Mac that has Final Cut Pro open.

There are some great advantages to shooting tapeless, as well as some disadvantages. In this chapter we will explore both.

For best results editing HD media in Final Cut Pro, use an Intel-based Mac, either the Mac Pro or MacBook Pro.

Sony XDCAM and XDCAM HD

When Sony introduced XDCAM in 2004, it also debuted a new system for managing huge SD and HD digital video files in Final Cut Pro. Every time you press its record button, an XDCAM camera makes two separate recordings of your content—the high-resolution clip you'll use for final output, and a low-resolution clip called a *proxy*. To simplify the workflow and expedite the process of working with XDCAM in both SD and HD, an editor can use the proxy files to preview and choose what portions of which clips are needed for the edit.

For instance, say you're editing down a 20-minute press conference clip for use in a 30-second sequence. Final Cut Pro imports the high-res XDCAM content roughly two to three times faster than real playback time, so transferring the entire high-resolution clip would take 7 to 10 minutes (not to mention filling several gigabytes of hard-disk space). If you work with the smaller proxy file instead and select, say, 7 or 8 seconds each of setup, wide shots, and cutaways, and then two 15-second sound bites, you'll wait just 20 to 30 seconds to import a minute of high-res content into FCP for final trimming and output. Using XDCAM proxy files can be a real timesaver.

Sony's standard definition XDCAM camcorders record onto high-capacity, DVD-like optical discs housed in rugged plastic enclosures. Recording time for each 23 GB XDCAM disc is about 85 minutes using Sony's standard definition DVCAM compression format at a capture rate of 25 megabits per second. On Sony SD cameras that support the IMX compression standard, XDCAM disc recording at the highest-quality capture rate of 50 Mbps consumes about 40 minutes, but longer recording times can be achieved by capturing with IMX at lower rates of 30 Mbps or 40 Mbps. Final Cut Pro natively supports all four of these SD formats in both NTSC and PAL.

In early 2006 Sony unveiled XDCAM HD, which captures in high definition and delivers compressed files to the same XDCAM discs at data rates similar to those for Sony's SD cameras and decks. XDCAM HD cameras can shoot in multiple SD and HD formats at a variety of capture rates, all of which, again, are supported natively in Final Cut Pro.

XDCAM supports three HD formats, which are called different things in the Sony world and in the Apple world. Before editing, you should collaborate with the photojournalist and standardize on one compression data rate. Here they are, with their Sony and Apple names:

Camera/Deck Setting	Final Cut Pro Setting
LP (long play)	XDCAM 35Mbit/s VBR
SP (standard play)	HDV
HQ (high quality)	XDCAM 35 Mbit/s VBR

The Sony XDCAM HD cameras can shoot 1080i HD video, but not 720p—an important consideration, depending on the network affiliation and broadcast format. In the United States, ABC and Fox broadcast at 720p, CBS and NBC broadcast at 1080i. For web delivery of HD podcasts, most likely 720p would be chosen to take best advantage of LCD and plasma screens and projectors.

Before anything is captured to an XDCAM disc, the disc must be formatted. If you put a brand-new XDCAM disc in a Sony XDCAM camera, the camera formats the disc automatically, using the settings applied to the last capture. You can't mix HD and SD on the same disc, but you can mix different HD compression settings on the same disc. To reformat a disc that has already been shot, use the menu controls on the camera or deck. (You cannot format an XDCAM disc using your Mac.) Formatting deletes all the files on a disc and prepares it for capture at the settings you specify. Formatting also resets the disc name to "untitled." You can rename the disc using the camera or deck menu, or you can rename once you mount the disc on the Mac. The hardware controls prompt you to confirm that you want to format the disc so you don't inadvertently destroy material you want to keep.

When the camera trigger is activated, it records many files to disc, which are all locked together and presented as one file. These files include the high-resolution video; the high-bit-rate audio tracks; the low-bit-rate, highly compressed proxy video/audio file mentioned earlier; a still picture of the clip's first frame (or *headframe*); and metadata files that describe all of these. You don't need to worry about these component files, as they are all linked together when displayed to you as the proxy in the XDCAM Transfer software, and the corresponding high-resolution images in Final Cut Pro.

> **NOTE** ► Metadata means "data about data." Using metadata can be invaluable in certain situations. For instance, a photojournalist may grab an unlabeled videotape and, thinking it must be new if it isn't labeled, shoot over what was an edited story that took two days to edit. Metadata can be clip names automatically added at capture, added fields of information describing the content of the shot, sound bite info, screen direction descriptions, and so on. Adding these descriptors provides databases with methods of searching and providing results based on that data.

Each clip appears as a single thumbnail when you view it in FCP.

If the photojournalist starts recording to disk 76 times, 76 separate shots are captured and then displayed as individual clip icons in the XDCAM Transfer interface. There's no need to scrub through the entire recording to find a shot. Once selected and transferred, these clips then show up in the Final Cut Pro bin along with metadata captured with the clip—in this case, clip name, camera serial number and model number, and date and time recorded, among others.

Initial Setup for XDCAM SD and HD

The Mac you are working on has likely already been configured to work with your XDCAM camera or deck, but you may occasionally need to configure a system from scratch. To do so, follow the steps in this section.

In order to edit, Sony XDCAM SD and HD devices require Sony XDCAM Transfer software to be installed on your system. In addition to the XDCAM Transfer utility application, this free download installs plug-ins and codecs for Final Cut Pro that seamlessly integrate file transfers between XDCAM Transfer and FCP. Sony's website offers versions 1.0 and 1.1 XDCAM Transfer software. You must install version 1.0 first, and then version 1.1 to get all the codecs loaded in the appropriate locations. You will also need to configure the XDCAM deck to be able to connect to your Mac Pro or MacBook Pro.

Installation of the Sony Software

Before you get started, open OS X Software Update (in the Apple menu) and install any updates to ensure you have the latest versions of OS X and Final Cut Pro. Then, download the Sony XDCAM Transfer software from the following URL:

https://servicesplus.us.sony.biz/sony-software.aspx?model=PDZKP1

Or do a Google search on *Sony XDCAM Transfer*. Find the PDZK-P1 download link and begin the download.

Quit Final Cut Pro prior to installation to ensure the proper placement of the various elements that get embedded in its interface. When the installation is complete, open Final Cut Pro and choose File > Import.

Confirm the presence of the Sony XDCAM menu option but do not select it. Its appearance means the plug-ins were properly installed into Final Cut Pro.

Once you have confirmed this, it is time to configure the deck.

Setting Up the XDCAM Device for the First Time

There are detailed setup instructions in the PDZK-P1 operation manual that downloads with the XDCAM Transfer software. If you don't have access to that, here is a short list of what to do and when to prep your hardware for use with Final Cut Pro:

1   Disconnect the FireWire cable from the deck or camera if connected.

2   Turn on the deck or camera and then press the Menu button.

3   Pressing the arrow keys, navigate to the menu and choose Enhanced (Basic won't work). Press the left arrow to navigate back to the home screen.

4   Go to System menu > setup menu > Operational, then scroll down to Interface.

5   From the Interface menu, choose i.Link Mode, then choose FAM (File Access Mode). Press Set.

6   Turn off the deck or camera.

7   Connect the FireWire cable from the deck or camera to the Mac.

8   Turn on the deck.

The XDCAM device's LCD display should read *PC Network*.

To disconnect from PC Network mode and return to normal operation using the XDCAM deck's shuttle knob and buttons on the deck, disconnect the FireWire cable from either the device or the Mac.

Software Setup for XDCAM Transfer

Follow the installer instructions. You will need to restart after installing 1.0, and then again after installing 1.1.

Print out the Sony XDCAM Transfer manual for easy reference. When you install the Sony XDCAM Transfer software, several plug-ins are installed into Final Cut Pro.

Before you open Final Cut Pro the first time after installing the Sony software, open the standalone XDCAM Transfer program from your Applications folder. Choose XDCAM Transfer > Preferences to specify where metadata and media will be delivered when you import the XDCAM media. The Preferences window that appears contains three tabbed buttons: General, Cache, and Import.

Click the General button.

Select this box and choose Final Cut Pro from the pop-up menu. A window will appear in the Sony transfer software that will also allow you to choose from several open projects.

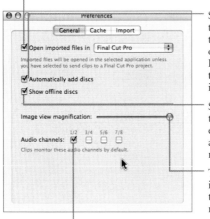

Select this box if you want the Sony transfer software to watch for new discs to be mounted and begin the transfer of their proxy files automatically. Leaving it unselected will force you to have to browse to each volume to initiate transfers manually.

Select this box to retain the view in the Sony transfer window for proxy elements that are "cached" (copied) after the XDCAM discs have been removed from the deck or camera.

This slider adjusts the proxy frame size in the Sony interface. This is much like the slider in Aperture and iPhoto for resizing your catalog.

If you recorded only two channels of audio, don't import more than that. Bringing in more clutters up the Timeline and also wastes disk space transferring "silence."

Now click the Cache button to target the transferred proxies to a specific drive or a server.

Select the folder in which you want the proxies to be delivered. If you're on a server-based system, your Media Manager will most likely set this up for you to deliver the media to shared storage for all to see. On a standalone Mac, the Movies folder is a good choice. You can screen proxies stored there using Front Row, and by synching that folder with iTunes, you can preview the material on AppleTV, if your system permits it. With the MacBook Pro, you can watch the proxies using Front Row if they are in the Movies folder.

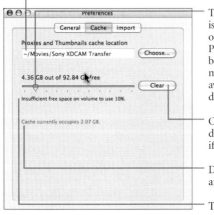

This slider allocates how much space is reserved for proxy transfers. When on a MacBook Pro or standalone Mac Pro, you will need to manage this space because you are going to be bringing in much larger files with a finite amount of available storage. Sony prefers 10% of the disk drive for proxy, but it works with less.

Click to clear the cache. A confirmation dialog will appear, so you needn't worry if you click it accidentally.

Displays how much of the cache you are using in that selected folder.

Thus, this warning.

Click the Import button to identify where you want the high-res material to be deposited.

Leave unselected. For the most part, material is going to stay on your system a short time. By overwriting you could disconnect media from an edited sequence by bringing it in again.

Click the Choose button and select the folder in which you want the high-res material to be delivered. On a standalone Mac Pro or MacBook Pro, you will want this to be in the Final Cut Pro Documents folder. If connected to Xsan, someone will have targeted the transfers to a specific shared volume.

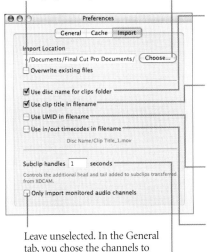

Select this. If you have discs with names (including "Untitled"), these names appear in the interface even when discs are ejected, allowing you to easily browse the proxy files.

Select this. It brings over the original recorded clip number. You can (and should) set the camera to use a specific file-naming convention. This setting is documented in the XDCAM camera manual.

UMID means "Unique Media ID." This is a *very* long character string and will absolutely clutter everything on your desktop. Leave this unselected.

Optional. If you don't set your cameras properly, every disc will start at C0001. If this is selected, the timecode start/end numbers will be added to the clip name in the bin.

Set this to 1 second. This brings in 1 extra second at the top and bottom of what you marked for import. A subclip is the portion of the original clip that you marked for import.

Leave unselected. In the General tab, you chose the channels to import. This option lets you change those channels directly in the user interface by toggling on and off the channels you are monitoring. To avoid errors when you're working under a time crunch, leave this unselected and just import all the recorded tracks.

On a standalone system, your target folder for the high-resolution material is named Sony XDCAM Transfer by default and placed in the Final Cut Pro Documents folder, itself located in your Home folder's Documents folder.

A typical installation, therefore, would transfer your high-resolution files to Macintosh Hard Disk/Users/*username*/Documents/Final Cut Pro Documents/Sony XDCAM Transfer. For a server-based installation with Xsan, a shared folder should be designated for you.

> **NOTE** ▶ The "Subclip handles" setting in the FCP import preferences extends the beginning and end of your clips by the designated amount of time, creating "handle" frames necessary for many transitions, such as dissolving up to your start frame or fading out from your final frame. Be aware, however, that when FCP adds handles to subclips at import time, it "forgets" the original In and Out points you designated, so you'll need to mark them again in the Viewer before applying transitions.

Once you have made these settings, confirm them each time you start working in a shared edit room or use a shared MacBook Pro. This will reduce confusion and eliminate panic when you're trying to find media under a deadline.

Software Setup for Final Cut Pro

The Final Cut Pro 6 Open Timeline lets you edit SD and HD material together in the Timeline. (This is explained in detail in Chapter 2, "Initial Setup.") You need to know, however, that in a file-based workflow using servers, in which you are delivering your story to a playout server as a file, you will need to render your sequence to a preselected format and compression. In this case, you will need to ensure your Easy Setup is set for the target format of your playout server.

In most cases, a station or network will decide on a preferred format and shoot and edit in that native format as a matter of policy. This is a great method because it reduces the confusion associated with changing formats and having different formats in which to edit.

Connecting an XDCAM Device to Your Mac

With the XDCAM deck or camera switched off, connect the FireWire cable between it and the Mac. Refer to the instructions earlier in this chapter to ensure that the deck or camera's i.Link options are set to FAM (File Access Mode).

Once it's connected, switch on the camera or deck.

Overview of the XDCAM Transfer User Interface

You can run the Sony XDCAM Transfer utility as a standalone application, or open it from within Final Cut Pro.

If you run the application by itself, you can use it to choose the clips and subclips you want from the proxy files, even if you're working on a Mac that is not your editing workstation. You can use XDCAM Transfer "offline" in a news truck, for instance, to select the elements to edit. When you mount the disc later, as a volume on an FCP-equipped Mac, your selections will be recognized and the appropriate high-res video clips will be transferred to FCP for editing.

While making offline selections with XDCAM Transfer can speed your workflow, all kinds of advantages are available if you make your selections on the same Mac you use to edit in Final Cut Pro.

In that case, open XDCAM Transfer from within Final Cut Pro.

To open from within FCP, choose Sony XDCAM from the File > Import menu.

Opening the XDCAM Transfer application from within Final Cut Pro automatically sets up transfers directly to the chosen project and bin. The Sony application can run concurrently with Final Cut Pro with no performance loss. In fact, while transfers are being made from the XDCAM disc, you can be editing clips that have already arrived in Final Cut Pro. There is no need to shut down the Sony XDCAM Transfer software while you are editing.

Let's look more closely at the proxy viewer and the process of selecting elements for editing. You can select portions of clips by marking them and importing them, or you can select entire clips by Command-clicking them in the proxy window and then pressing the Option key and clicking the Import All button. (Pressing Option switches the Import option to Import All.)

Log and Transfer from Tapeless Media

An XDCAM disc connected through a deck or camera will mount and appear here, above the offline discs.

Name of disc and number of remaining proxies on disc.

Name of disc from which proxies are displayed.

Entire clips can be selected and imported without their being logged. Command-click lets you select non-adjacent clips.

More offline discs displayed with number of clips for each disc. Note the importance of naming your disks to avoid "Untitled" confusion. *Offline* indicates the disc is not mounted or inserted in the deck.

Transport controls for Viewer.

Mark In/Out buttons. Also displays marked timecodes and durations.

Proxy selection window. When a disc is mounted for the first time, these will appear blank with progress bars as the proxies arrive sequentially, oldest first. You may interrupt and scroll down to the last shot to force that proxy to display while the others are arriving.

If you select more than one segment of an individual proxy clip (the press conference mentioned earlier in the chapter, for instance), and import them as subclips, Final Cut Pro assigns each subclip's name an incrementally numbered suffix. The first subclip from clip C0005 is named C0005_1, and so on. This lets you quickly see what components of a single clip were imported to Final Cut Pro.

You may view clips in the XDCAM Transfer proxy-selection window in a sortable list rather than a series of thumbnail icons. Choose View > as List. This is a great way to sort clips using timecode values.

32 Log and Transfer from Tapeless Media

Columns may be sorted by clicking the column header, as well as rearranged by dragging and dropping. For instance, if you want the newest clip at the top, click Name or Start since the clip name and timecode are sequential.

Anatomy of an XDCAM Transfer Clip Proxy Icon

The thumbnails in the XDCAM Transfer proxy-selection window provide a lot of at-a-glance information about the clips they represent.

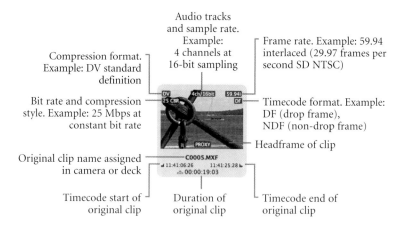

Mounting the XDCAM Disc

Do a test of the process of mounting the XDCAM disc and absorbing the media to confirm it all works before you have to perform under deadline.

1   Connect your Mac and the camera or deck with a FireWire cable and, with Final Cut Pro and XDCAM Transfer shut down, insert the disc in the camera or deck to mount it on the desktop.

2   Rename the disc (change it from "Untitled") to make finding things easier later. This can be done on the desktop by clicking in the name of the disc (Untitled), or Control-clicking the disc icon, selecting Get Info from the shortcut menu, and then renaming the disc in the window that appears.

    There is no current way to rename the disc directly within the Sony XDCAM Transfer interface. It must be done at the desktop level or, better, in the camera before it gets to the edit room or truck. Also, reformatting the disc in the camera or deck erases the disc name, resetting it to "Untitled."

    NOTE ▶ Make sure the disc's "protect" tab is disabled. This is a tab on the left side of the thin spine of the disc. When you insert the disc in the deck, the spine is the part that faces you before it disappears inside the deck or camera. If the disc is protected, you will see a red marker. To unlock it, slide the tab to the right. If the disc is in write-protect mode, you will be unable to change its name.

3   Open Final Cut Pro and choose File > Import > Sony XDCAM to open the XDCAM Transfer software.

    The Sony window opens and proxies begin transferring to the selected cache folder.

4   Select a shot by clicking its thumbnail in the XDCAM Transfer proxy-selection window. The shot's first frame will appear in the XDCAM Transfer Viewer.

5   Mark In and Out (I and O) points on that clip and click the Add (+) button in the window, or press Return to add the selected subclip to the import list. Mark more elements of that clip as desired, and then click the Import button to transfer subclips to FCP.

A progress bar appears in the proxy thumbnail for each clip or subclip as its high-resolution content transfers into Final Cut Pro. Once a clip's progress bar is complete, you can begin editing it in FCP.

6   Open the clip in the FCP bin and play it, or drag it to the Timeline to edit it. Only after importing a clip to FCP can you rename it, add information to columns, add markers, and so on.

7   Repeat steps 4 through 6 as necessary.

If Good Shot Marks were added while shooting or screening in the camera or deck prior to mounting on FCP, these marks display as FCP Markers in the Viewer and Timeline.

> **NOTE** ▶ Nothing will stop you from marking and reimporting clips you may have already marked and imported. Importing clips more than once uses up more storage as well as wasting time.

You may also choose to make your clip selections as one procedure, and then import all the selections at once. To do this, press the Option key to change the Import option to Import All and then click the button.

Using the Sony XDCAM Transfer Software Within Final Cut Pro

Once you mount an XDCAM disc, a window appears to the right of the disc icon that tells you how many clips are on that disc. The icon looks like an XDCAM disc with a number in it. Transfer of the disc's proxy files begins immediately. You will see the outlines of each proxy thumbnail first, and then gradually the headframes or thumbnails will appear.

By default, proxies transfer in the order they were shot. If you need to work right away with a shot other than the first one that was taken, you can scroll down to the proxy you need and then click it. The system will interrupt the sequential transfer process and reveal the selected proxy in the Viewer. You can build a selection list for this clip, and you can interrupt the proxy transfer to load the high-res material into Final Cut Pro immediately.

Make sure you have selected the correct bin in the "Send clip to Final Cut Pro project" pop-up, and that you have selected the box next to its name.

This will ensure delivery of the clip to that specific bin (whatever project you named it).

As you begin selecting elements from within the individual clips, use the small Mark In button on the window (or press I) to select the start frames for your subclips, and Mark Out (press O) for the end frames. Then click the Add (+) button or press Return to add the element to your import list, which is on the bottom right of the window.

You can control the Viewer transport controls with keyboard shortcuts as well. As in Final Cut Pro, the spacebar will either play or stop a clip. The J key lets you move backward, and the L key lets you move forward at normal play speed. If you press JJJ, you will be playing backward at around four times the normal speed. If you press the spacebar or K, the clip will stop. If you press LLLL it will go forward (with audio) at about eight times the speed. If you press J, it slows down and goes only at four times the speed. Unlike FCP, the Sony software does not enable you to press JLJL to scrub over a point.

Once you finish adding elements from that particular clip to your selection list, click Import. The selected high-res material begins transferring into the Final Cut Pro bin.

Please keep in mind that if you don't click Import after having made selections, and you go to another clip, your selected content will not be transferred. If you inadvertently move on to a new proxy without clicking Import, you can click back to the skipped clip, and its subclip list should be there.

Eject the Disc

Once you have completed the selection process and have transferred your selects to Final Cut Pro, you can eject the disc. There are several ways to do this, and the most obvious of these is very dangerous: *Do not, under any circumstances, use the Eject button on the deck or camera while it is connected to Final Cut Pro.* Instead:

▸ From the XDCAM Transfer interface, highlight the disc in the top left of the Source window, then click Eject at the bottom of that window.

▸ Highlight the disc in the top left of the Source window of the interface, and choose Eject from the File menu.

▸ From the Macintosh desktop, Control-click the disc icon and choose Eject from the shortcut menu.

▸ Drag the disc icon from the Mac desktop to the Trash.

Since the disc is a volume with file structures that are being accessed, you must eject it "gracefully." Although the deck is in PC Network mode, its physical Eject button is still active. Again, pressing the Eject button on the deck or camera is *not* graceful and should be avoided.

To release the deck or camera from the PC Network mode, you must physically unplug the FireWire cable from either the Mac or the deck or camera. Once the cable is unplugged, the deck or camera can be used in normal operational mode. If you reconnect the FireWire cable, it will go back to PC Network mode.

Panasonic P2

The media used in Panasonic P2 cameras have no moving parts, which makes them impervious to the mechanical wear and tear that can thwart videotape and some hard disk systems, especially under punishing conditions, such as in aerial deployments or dusty or humid field conditions.

P2 media resemble old laptop modem or Ethernet cards, and in fact fit into the same PC Card or CardBus slots as those devices. Apple's PowerBook laptops had PC Card slots, but the modern Macs providing a higher level of performance with FCP 6 do not, so unless you are connecting directly to a P2 camera, an adapter is required to mount P2 media directly on a Mac Pro or MacBook Pro.

The P2 card is a very high-performance cousin to the SD media you may be using in your digital still cameras, GPS devices, and PDAs. Each P2 card contains four high-speed memory modules, and data are "striped" across all four for greater bandwidth. Striping, or writing data across multiple storage devices at the same time, is also used to record uncompressed HD onto disk drives.

P2 cards are available in several capacities. The total recording time each allows is governed by its capacity and your choice of compression settings. Cards are available in 2 GB, 4 GB, and 8 GB, with 16 GB cards just around the corner.

- 2 GB at DVCPRO 25 will yield almost 10 minutes.
- 4 GB at DVCPRO 25 will yield almost 20 minutes, at DVCPRO HD up to 4 minutes.
- 8 GB at DVCPRO 25 will yield almost 40 minutes, at DVCPRO HD up to 8 minutes.

You may recall that back in the days of film-based news gathering, 100 feet of film, or a little over three minutes' recording time, was standard for a story, and film had to be processed for an hour before it could be edited. Those recording durations were certainly manageable back then, but since the advent of lower-cost videotape in the mid-1970s, news and sports crews have shot more liberally. Filling a 20-minute tape cassette was typical per story.

Panasonic realized that P2's low per-card capacity would require multiple cards per cameras. Its pro broadcast cameras have five P2 card slots. The smaller "prosumer" camera has two slots. This allows you to continue recording across more than one card and stitch the elements together in the edit room.

All of the DVCPRO compression formats can be edited natively in Final Cut Pro, including HD at 720p and 1080i at various frame rates. They require no conversion of their actual video content, or "essence." Since the Panasonic P2 cameras can shoot in 720p and 1080i formats, they can be used natively at the ABC, Fox, CBS, and NBC networks and their affiliates.

The simplicity of the interface provides a fast and easy method for editing material from the P2 cards. However, since the file format on the card is not a native QuickTime movie, one step needs to be taken for Final Cut Pro to effectively "rewrap" the files as QuickTime for editing in the Timeline.

Importing P2 Content into Final Cut Pro

If you are using the five-slot P2 deck, connect its USB 2.0 or FireWire cable to the Mac. The P2 deck requires AC power and is not "bus" powered from the computer. You may also connect the camera to mount the cards and transfer the material.

Once the camera or deck is mounted, you will see the icon(s) of its cards on the Mac desktop and in the folder view, as shown.

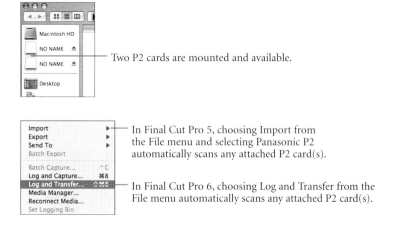

Two P2 cards are mounted and available.

In Final Cut Pro 5, choosing Import from the File menu and selecting Panasonic P2 automatically scans any attached P2 card(s).

In Final Cut Pro 6, choosing Log and Transfer from the File menu automatically scans any attached P2 card(s).

Previewing Shots

The P2 Log and Transfer window opens, embedded within Final Cut Pro, and immediately displays the material stored on the mounted card(s).

Each clip's thumbnail displays immediately: no waiting. Each thumbnail is a still image captured to the P2 card, along with the video and audio files.

When the shot you've selected from the list appears in the Viewer, you can work with it using standard Final Cut Pro controls and keyboard shortcuts.

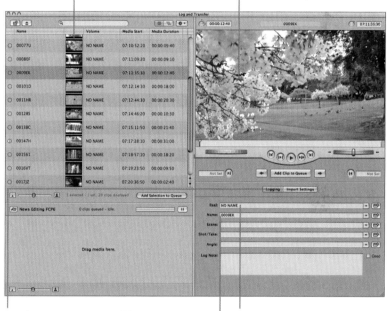

Use the Logging pane to add or change metadata for each clip before transferring it into Final Cut Pro. Wise use of even a few of these fields can prevent confusion and keep the editing process well organized.

Reel name defaults to No Name unless you've specified a different default on the camera.

Clip name is automatically assigned when each shot is started. You may rename each shot here prior to import and delivery to Final Cut Pro.

You can click in a desired shot, scrub through it, and mark In and Out points from the P2 card directly to make that selection available in the Final Cut Pro bin. You can also build a list of selections and transfer them all at once. You can shoot at different compression settings on the same card (25, 50, or 100). Editing and playing all of these in the Timeline is available in Final Cut Pro 6, but if you are playing out of a server it is suggested to conform the sequence to the target compression and frame rate of the server to avoid unnecessary rendering.

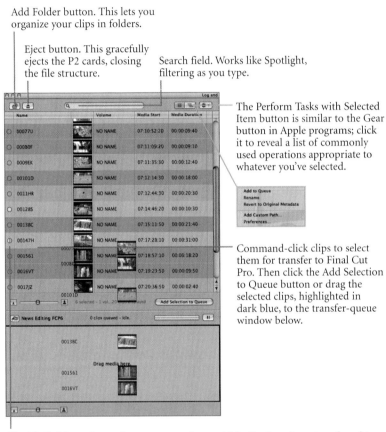

Add Folder button. This lets you organize your clips in folders.

Eject button. This gracefully ejects the P2 cards, closing the file structure.

Search field. Works like Spotlight, filtering as you type.

The Perform Tasks with Selected Item button is similar to the Gear button in Apple programs; click it to reveal a list of commonly used operations appropriate to whatever you've selected.

Command-click clips to select them for transfer to Final Cut Pro. Then click the Add Selection to Queue button or drag the selected clips, highlighted in dark blue, to the transfer-queue window below.

The Media Map column shows you at a glance which clips have been transferred to Final Cut Pro. Blue circles indicate clips transferred in their entirety. Unshaded circles denote clips that haven't been transferred. A partially shaded circle indicates that one or more portion(s) of a clip have been transferred as subclip(s) to FCP.

Panasonic P2 41

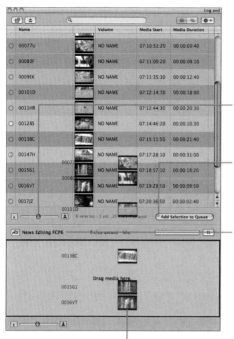

Slider adjusts size of the P2 clip thumbnails, much like similar controls in Apple Aperture and iPhoto.

If you want to import all the clips on the card, click in the list and the press Command-A to highlight all, then click the "Add Selection to Queue" button.

The target bin's name is displayed here. To switch bins, Control-click the new target bin's icon in FCP's Browser and choose Set Logging Bin.

Highlighted clips are dragged here and transferred as marked, or in their entirety if not marked.

Clips placed in the Transfer Queue window transfer one at a time into Final Cut Pro, topmost first. You can drag clips up or down in the queue to change the transfer order. Once a clip's transfer is complete, it is available for editing in FCP. The clip disappears from the queue window (but remains in the main clip list above).

Spinning icon indicates the clip currently being transferred.

Progress bar of the transfer for a particular clip.

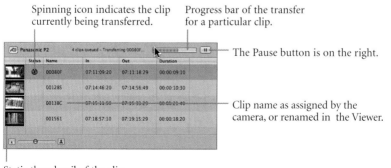

The Pause button is on the right.

Clip name as assigned by the camera, or renamed in the Viewer.

Static thumbnail of the clip.

Click the Import Settings button to reveal controls for specifying which elements of a clip you want to import.

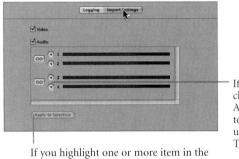

If you didn't capture on channels 3 and 4, click the Audio Capture Channel buttons to deselect them. This speeds up the transfer and reduces Timeline clutter when editing.

If you highlight one or more item in the clip list window before adjusting any of these settings, the Apply to Selection button activates. Click it to apply changes only to the selected clip(s). Otherwise, changes made here apply globally to all clip transfers.

Anatomy of a P2 Viewer

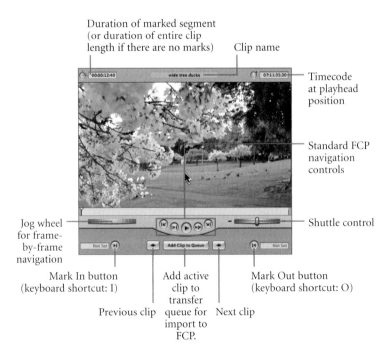

Duration of marked segment (or duration of entire clip length if there are no marks)

Clip name

Timecode at playhead position

Standard FCP navigation controls

Shuttle control

Jog wheel for frame-by-frame navigation

Mark In button (keyboard shortcut: I)

Add active clip to transfer queue for import to FCP.

Previous clip

Next clip

Mark Out button (keyboard shortcut: O)

As the material is being delivered into Final Cut Pro, you may begin editing with it. The clips appear in the destination bin and are ready to be dragged to the Timeline or opened in the Viewer.

Ejecting the P2 Card

There are several ways to eject the P2 card or cards:

▶ Click the Eject button in the P2 window within Final Cut Pro.

▶ Drag the P2 card icon from the desktop to the Trash.

▶ Select and eject the P2 card using any other standard OS X methods for ejecting volumes in the Finder.

Ikegami Editcam

The first-ever tapeless field acquisition device, the Ikegami Editcam, dates back to 1995. It has evolved into a line of SD and HD cameras, which record onto FieldPak storage modules that can consist of hard disks or static RAM modules similar to those used by Panasonic.

As with P2 and XDCAM, the Editcam file format is not QuickTime native, so it is not possible to connect an Ikegami FieldPak directly to Final Cut Pro and begin editing. There needs to be software in the middle to unwrap the file and rewrap it as QuickTime.

Telestream's Flip4Mac Editcam software provides this capability. Currently Flip4Mac Editcam supports only standard definition compression such as DV-25, DV-50, and IMX resolutions.

To edit Editcam content, first mount the FieldPak as a volume, using an NL Technology SAT-110 reader connected to your Mac via USB 2.0. Insert the FieldPak into the reader and it will mount on the desktop.

Open Final Cut Pro and choose File > Import > Ikegami Editcam to open Flip4Mac Editcam.

Choose this if you are connected to a server or other hard drive with other standard definition MXF files to import.

Choose this if you are connected directly to a FieldPak with the SAT-110 device.

Flip4Mac Editcam will seek out all the clips on the FieldPak and display them as thumbnails. Select the clips you want to use and click Import. These clips are then deposited to the Final Cut Pro bin you specify in the Flip4Mac preferences. (You can preview the clips from within Flip4Mac Editcam, but you currently cannot mark or select specific clip segments for importation; you must import each selected clip in its entirety and make edits in Final Cut Pro.)

Check boxes in this column to select individual clips for import.

Choose the Capture Scratch folder for FCP as the destination.

Viewer with basic navigation controls

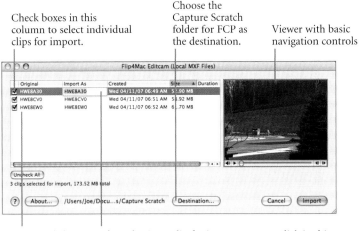

Original clip name from camera here.

After selecting a clip for import, you can click in this column and type a new name that will be assigned to the file when it is imported to Final Cut Pro.

Select a clip you want to import by clicking the box to the left of the original clip name. This activates a text field in the "Import As" column, where you can (and should) assign the file a more convenient name for use in Final Cut Pro.

In order for Flip4Mac Editcam to deliver the material directly into your active bin, you must designate the destination in the Flip4Mac user interface. To do this, click the Destination button and navigate to the desired destination folder.

The Capture Scratch folder that Final Cut Pro creates in your Home/Documents folder is a good destination to use. It is wise to get in the habit of selecting it every time you start a session with Flip4Mac Editcam. If you skip this step, Flip4Mac Editcam will drop your media files in whichever project bin you happened to open first after launching FCP—nonintuitive behavior that can lead to needless searching for files on deadline.

On the lower right of the dialog, click the Import button. The selected clips will be deposited to the Final Cut Pro bin you specified.

Ejecting an Editcam FieldPak

To edit an Editcam FieldPak, close the Flip4Mac Editcam application and drag the disk icon from the desktop to the Trash. The Eject button becomes active on the SAT-110 only after the file structure is secure and ready to eject.

Focus Enhancements FireStore

The FireStore is a unique media-capture device in that it can be set to record native QuickTime in both SD and HD formats. Connecting a videotape camera's FireWire port to the FireStore lets you record onto tape and disk at the same time. The FireStore will start and stop recording at the same time as the tape. This is a great way to shoot one copy for editing (FireStore) and one for the shelf (tape).

When you finish shooting, connect the FireStore directly to your Mac's FireWire port to mount it as a desktop volume.

Video material on the FireStore volume is available for editing directly from within Final Cut Pro. You can edit in HD if you captured HD, and in SD if you captured in SD.

This process is great for late-breaking news and sports content, as it eliminates all need to "rewrap" media or import it into Final Cut Pro.

Ejecting a FireStore Drive

To eject a FireStore drive, drag the drive icon from the desktop to the Trash. Disconnect the FireWire cable.

4
Log and Capture from Videotape

Most broadcast organizations aspire to tapeless acquisition, but the reality is that most are still using videotape in the field. This chapter attempts to accelerate the process of working with videotape in a nonlinear environment, specifically as it relates to fast-paced news and sports editing.

The chief difference between transfers from videotape and tapeless systems into Final Cut Pro is that videotape content can be captured to a hard disk or server only in real time—importing a 15-minute tape clip takes 15 minutes—while tapeless media allow faster-than-real-time importing. (At one time Sony made a high-speed DVCAM tape deck and Panasonic made a high-speed DVCPRO deck that offered a faster-than-real-time ingest method called Serial Digital Transport Interface, or SDTI. Those decks are rare and are no longer manufactured.)

Many broadcast stations have designated "ingest" stations, where new videotapes are captured to servers in real time as soon as they arrive at the station. This method provides access to raw material from all networked editing seats while the material is arriving on the server.

Ingesting the full contents of every inbound tape works well for many applications, but it falls short when, for instance, five photojournalists arrive back at the station with breaking-news footage just 20 minutes before airtime. This is where capturing directly in the edit room accelerates the process.

If you share the edit room (or a laptop) with other editors, it is always a good idea to confirm that the Easy Setup menu is set to your capture and delivery format before editing. This will maximize the performance in real time, as well as expedite the transfer to a playout server that is set to the same target format and codec.

The two fundamental rules for accelerating a videotape workflow are as follows:

▶ Don't log first and then capture if you're under a deadline. Instead use the Capture Now button and grab material as you screen it.

▶ Don't mark In points prior to capturing. It wastes valuable time because the deck will have to "pre-roll." Again, use the Capture Now button to grab material on-the-fly.

After a couple of times, you will remember all the details of the setup and capture, even under deadline pressure.

Logging Videotape Prior to Capture

Logging is the process of identifying material using timecode to describe certain elements within a shot. This is done either on paper or on an electronic log that can be used to retrieve video content automatically. Typically this process is used for episodic television, movies, commercials, documentaries, and longer-form news segments when multiple takes and alternate shots are available. Producers of breaking-news and sports segments seldom have those luxuries, or the time to sift among them, so this guide does not cover the logging process.

Screening material, making selections, but *not* capturing that material at the same time when editing a news story is an impractical use of time. But if producers or reporters want to take timecode notes on sound bites, B-roll, or standup takes prior to walking into the edit room, let them. It is easier to have them hand you a sheet with timecode hit times and to load those elements directly than going through what is effectively an "offline" capture for a news story.

Preparing to Capture from Videotape

Whether you are connected to a server or working with standalone storage, the process of capturing is the same: only the destination of the captured media changes. The destination can be chosen in the Capture Settings tab in the Log and Capture window.

To open the Log and Capture window, choose File > Log and Capture, or press Command-8.

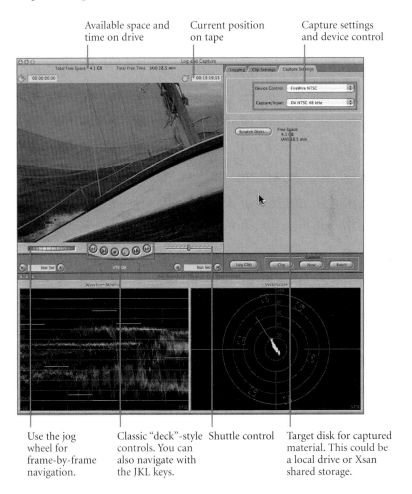

Available space and time on drive
Current position on tape
Capture settings and device control

Use the jog wheel for frame-by-frame navigation.

Classic "deck"-style controls. You can also navigate with the JKL keys.

Shuttle control

Target disk for captured material. This could be a local drive or Xsan shared storage.

Log and Capture from Videotape

Mark In and Timecode display. Don't mark an In point, because it will require the deck to go into a "pre-roll," wasting valuable time.

Mark Out and Timecode display. You can ignore this when doing direct capture from videotape (without pre-roll or logging).

Offline logging workflow

Capture choices: Clip, Now, and Batch

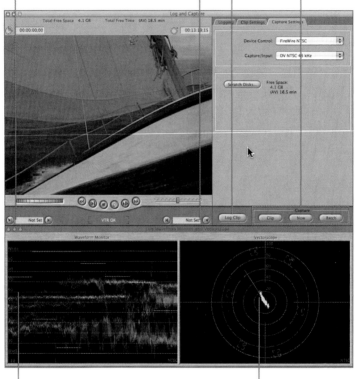

The Waveform Monitor represents the value of the brightness or "luminance" of the image. The closer the display gets to 100, the brighter the image is onscreen. You will not want to go over 100.

The Vectorscope displays the color value, or chrominance, of the image. The farther out from center, the more color is represented. For instance, a black-and-white image would appear as a point in the center. Six boxes on the display represent the six main colors of color bars.

The Capture Settings tab, in the upper-right corner of the Log and Capture window, contains several controls that must be set correctly for successful capture from videotape.

Begin with the Device Control pop-up, which enables Final Cut Pro to start, stop, and otherwise control your deck via its FireWire connection (or its connection to a FireWire converter, if the device itself doesn't support FireWire), as well as a 9-pin (RS-422) serial deck connection (either provided through a third-party capture card, or a USB adapter).

Once you have specified the correct Device Control settings, you should use the Capture/Input pop-up to set the frame rate and compression settings that will be applied to the tape content as it is imported into Final Cut Pro.

If you share your system, you should confirm your settings are correct before starting a project with a close deadline.

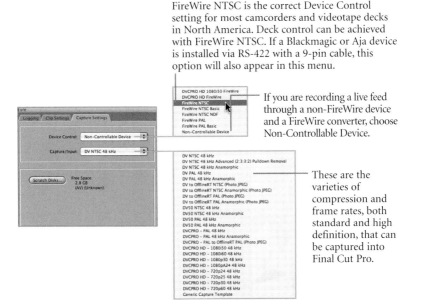

FireWire NTSC is the correct Device Control setting for most camcorders and videotape decks in North America. Deck control can be achieved with FireWire NTSC. If a Blackmagic or Aja device is installed via RS-422 with a 9-pin cable, this option will also appear in this menu.

If you are recording a live feed through a non-FireWire device and a FireWire converter, choose Non-Controllable Device.

These are the varieties of compression and frame rates, both standard and high definition, that can be captured into Final Cut Pro.

Click the Clip Settings tab. This is where you will select the components of the tape you want to capture, such as audio and video.

It's generally wise to capture the video and both audio tracks from all field tapes. You can always delete unnecessary or unused elements, such as audio from alternate mics, in the Timeline. It is much more difficult to go back after the fact to capture a track you missed—as when, for instance, a reporter records interview sound on the wireless-mic track while you're shooting coverage in a different direction.

This toggles the visibility of the Waveform Monitor and Vectorscope.

Video capture checkbox. If this is unselected, no video will be captured.

Audio capture checkbox. If this is unselected, no audio will be captured.

Input Channels pop-up. No reason to capture four channels if only two were recorded in the field.

Audio level meters

Stereo/Mono toggle control

Capture Audio Channel toggle controls

Last, click on the Logging tab. We are going to use this to name the clip, not to log clips offline and then batch capture.

This tab lets you set the logging names for the clips as well as the target bin. You may want clips to be named sequentially as you grab the shots you want: for instance, *exterior, exterior-1, exterior-2,* and so on. The naming convention can be set in this tab.

Note that you cannot name a clip while it is being captured, so you can either just let the naming happen by itself (number suffixes increment automatically), name the clips after they've been captured (the current best practice), or name them prior to clicking the Now button.

Preparing to Capture from Videotape   53

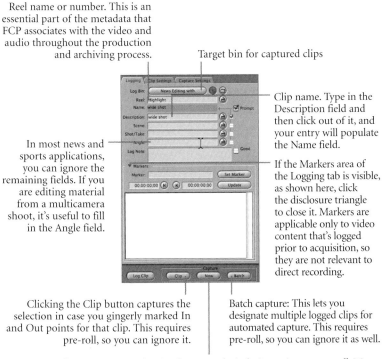

Reel name or number. This is an essential part of the metadata that FCP associates with the video and audio throughout the production and archiving process.

Target bin for captured clips

Clip name. Type in the Description field and then click out of it, and your entry will populate the Name field.

In most news and sports applications, you can ignore the remaining fields. If you are editing material from a multicamera shoot, it's useful to fill in the Angle field.

If the Markers area of the Logging tab is visible, as shown here, click the disclosure triangle to close it. Markers are applicable only to video content that's logged prior to acquisition, so they are not relevant to direct recording.

Clicking the Clip button captures the selection in case you gingerly marked In and Out points for that clip. This requires pre-roll, so you can ignore it.

Batch capture: This lets you designate multiple logged clips for automated capture. This requires pre-roll, so you can ignore it as well.

For direct capture, use the Now button exclusively. It requires no pre-roll. Map this command to a keyboard location such as F1 (see Chapter 10, "Customization Overview"), so you can shuttle to a desired starting point on the tape, play the tape, and then press F1 to begin capturing, and Esc to stop capturing.

Pre-roll during capture is unnecessary and should be avoided at all costs, but thousands continue to do it, even in news. It is a holdover from the linear tape-to-tape world, in which you had to back your decks up at least 5 seconds from the start of a selection, pause, and then play to make sure that both the player and recorder were synced prior to capture. The net loss for this outmoded practice is anywhere from 6 to 10 seconds per shot. Ten shots equals at least one minute lost, and you don't have a minute to lose.

Instead of pre-rolling, Final Cut Pro lets you set your deck to All Enable mode, so that you can control it using either its physical buttons or the transport controls in the FCP interface. Shuttling using the deck's knob or FCP's slider, for instance, has the same effect. Once you set the deck to All Enable you don't have to set it again. Most Sony Betacam decks have menu setting 006 assigned to this. Panasonic uses menu setting 004 for most of its decks. These menus are adjusted using either the front panel display on the deck itself, or using a monitor attached to the TC Char output, which is usually output 3.

Once this is all set up, you can begin the capture of the material from tape.

Capturing from Videotape

We are going to avoid searching around on the tape and grabbing things in the order we may want them edited. Instead we are going to re-rack the tape to the top and grab the elements we want effectively "live" into the bin.

> **NOTE** ▶ If you're working on a laptop, reset the use of your keyboard's function keys to control software. Open the Keyboard & Mouse System Preferences window and select the appropriate checkbox. You will then need to use the Fn (Function) key to adjust the screen brightness and the audio levels with the assigned function keys.
>
> ☑ Use the F1–F12 keys to control software features
>   When this option is selected, press the Fn key to use the
>   F1–F12 keys to control hardware features.

You will need only two commands in Final Cut Pro to accomplish a fast capture. Click the Now button in the Log and Capture window to begin recording, and press the Esc (Escape) key to stop recording and prepare for the next clip.

The default keyboard shortcut for the Capture Now button is Shift-C, but you should consider changing it to a single-keystroke shortcut such as the F1 function key, just to the right of the Esc key. Key remapping is described in detail in Chapter 10, "Customization

Overview." Remapping that shortcut will greatly accelerate the capture process and have you editing that much faster.

Here's a rundown on the steps:

1 Start shuttling through the tape from the top on the deck or with the jog wheel. When you get to the first shot you want, shuttle back a little. See the note on the following page about shuttling mini DV tapes.

2 Click the Play button. **3** Click the Now capture button (or press F1 on the keyboard if you've remapped it) to begin capturing the content. Wait a couple of seconds after you see the last part of the shot you want to capture.

When you're done, press Esc to stop the capture.

If you have timecodes for sound bites or certain NATSOT (natural sound on tape) hits, shuttle to just before those points. Click the Play button and then click the Now button (or press F1 if it's remapped). Play through the bite or NATSOT, and then press Esc to stop.

Give yourself a few seconds prior to the sound bite or NATSOT, as you may have to backtime into the sound later. It's better to have grabbed the sound once than to have to go back and get it later. This facilitates easy backtiming and extending sound bites into B-roll when you don't have time to find cutaways.

NOTE ▶ If you are using mini DV tape with a consumer camcorder, you should never shuttle the tape either from the FCP interface or the camera itself. These tiny wonders of mechanical achievement were designed to record video only, and using them to view tape is to be considered a luxury. Their motors and gears are intended to work at play speed, which is 1x forward. You risk any and all material on your tape as soon as you begin shuttling backward, as the thickness of these tapes (or thinness, as the case may be) can cause the tape to stretch or break as the gears shift from forward to backward. Some cameras are better than others, but based on experience, if you have to back up a mini DV tape, press Stop, then Rewind, then Play.

Having selected All Enable from the deck's setup menu, you can type timecode values for your clips' start points and use them for your capture, rather than shuttling to them. If you choose to enter timecode values in the Mark In window, you can press the Go To In button next to it or press Shift-I. Keep in mind that entering In points will require a pre-roll. In some cases the reporters may have spent more time identifying and transcribing the sound bite than in noticing the correct timecode. Therefore it is a good idea to subtract 5 seconds or so from reporter-logged values, just in case.

Still another direct-capture method involves the use of third-party software such as Gallery PictureReady, which enables FCP to automatically capture whatever you play using the deck or camcorder's physical transport controls. The PictureReady software "watches" for the sync signal that comes off the deck when it is in play mode, and tells FCP to capture only what you play. This can be a tremendous timesaver and helps to overcome the obvious drawbacks of using videotape this far into the 21st century.

Once your clips are captured, you are ready to lay out your story in any of three formats: VO (voiceover), VOSOT (voiceover/sound on tape), or package.

5
Viewing Media with Final Cut Server

Final Cut Server (FCS) is a server application with a Java client that runs on both Mac OS X and Windows. This application adds media-management capabilities to the Final Cut Pro workflow. It can be configured to work in shared-storage environments or on a single Mac server with directly attached storage.

Final Cut Server gives producers, writers, and reporters who don't have Final Cut Pro installed on their computers access to low-resolution proxy files stored on networked storage devices on-site, or access to media located in remote offices or bureaus. FCS lets these journalists browse stored content and select material for use in their stories. FCS users can also make simple rough edits, isolating the portions of video they want to use from long news conference or interview clips, for example. FCS users can organize and arrange these "subclips" into storytelling order and add notes or comments describing their requirements before forwarding the sequence to the editor who will prepare the final project for delivery using Final Cut Pro. This chapter provides a brief overview of the interface and functionality of FCS. (For a deeper understanding of the capabilities of Final Cut Server, please refer to the *Apple Pro Training Series: Final Cut Server Quick-Reference Guide,* by Matthew Geller.)

The Producer or Reporter's Workflow

Producers or reporters start off by opening the Final Cut Server application on a Mac or PC and doing a search for material they need for a story. As they screen the results of the search, they can decide to include material in what is effectively a folder. This folder, or *Production*, becomes the reservoir for any accumulation of shots (such as video, audio, and graphic elements) that may be selected for the story.

The Assets window displays thumbnails of all available clips.

The Productions tab is used to collect related assets in one place.

Use the search field to locate and isolate specific content.

Dragging selected clips here creates a new project. The FCP Create dialog opens to let you enter pertinent descriptions of the material so that it can be searched by others.

Clicking the "create new Final Cut Pro project" button from the toolbar displays the FCP Create dialog.

Settings for Destination and Job Priority can be set in case a file conversion is necessary. In this case No Conversion is chosen to preserve the source clips in the format in which they were ingested.

Because Final Cut Server is a media manager, information describing the material must be entered so that searches can locate material selectively. No info, no results. Anything new that you have added to the project can have specific metadata (data about data) automatically inserted into it.

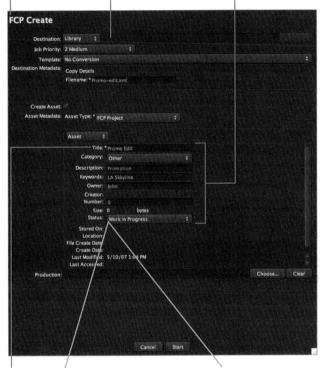

The producer or writer identifies the contents of the project so the editor can find the material.

Status pop-up. Final Cut Server allows changes in a project's status setting to trigger automatic alerts, notifying team members that a project requires their attention. For instance, if a reporter changes the status from Work in Progress to Rough Cut, an editor might be notified that the project is ready to cut together. These notifications are set up on a site basis and are customizable for the workflow of a particular facility.

The shot selection tool allows you to assemble simple rough-cut sequences. This allows a producer or reporter to screen sound bites and press conferences, for instance, to isolate the shots of interest.

Double-clicking a new FCP project opens this detailed view of the project.

Customizable button bar

The various methods of gathering, restricting, and reviewing material can be chosen here.

Producers or writers can use the shot selection tool to make rough-cut edits on the low-res proxies. If they are editing more than one clip, Final Cut Server will automatically string the edits together to provide a low-res preview of the edits before they are passed on to the editor.

Use the Source window to play or shuttle through uncut source clips loaded in the project, and to select elements from the clip by marking In and Out points.

The Preview window lets you play back a proxy view of the edited shots you've created, listed below. Click a clip name in the list to display it in this window.

The Producer or Reporter's Workflow   61

Standard transport controls

Thumbnails for source clips (uncut clips loaded into the project). Click one to display it in the Source window above.

Slider lets you drag through the clip or click directly to snap to position.

As you set each In and Out point in the Source window, those edited selections are displayed in this list.

Save the edited clip selection as a sequence in your FCP project.

Mark In and Out points on the areas needing annotation.

As annotations are added, they appear in a list here.

You can add or attach comments to clips in the Annotate window, which is a good way for producers to isolate sound bites and other interesting material.

Once the project is complete, simply saving it finishes the organization part of the editing process. The "pointers" to the selected low-resolution elements are now saved and available for others in the newsroom.

The Editor's Workflow

Final Cut Server lets you customize the notification process so that people involved in production can be alerted when it's their turn to add their part to the story. In this case, the editor is notified via email that a new project is ready for editing, then opens Final Cut Server and searches for that project. The editor is able to search using anything that's in the project metadata, which drastically cuts down on wasted time.

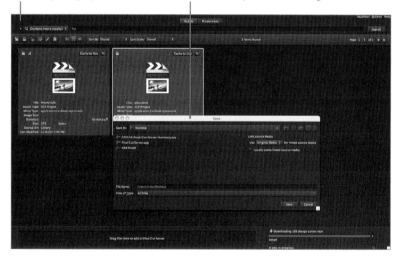

This Assets window displays the results of a search on "fcp," showing two projects.

This is the Save dialog. An editor can check out a project to work on within Final Cut Pro. Opening the project in Final Cut Pro will automatically link it to the high-res material.

The editor checks the project out and opens it in Final Cut Pro. The rough-cut edits the producer made will automatically link to the high-res material if both editor and producer are working on a SAN, or the media itself can be loaded on the FCP machine if they're working over Ethernet or a remote connection.

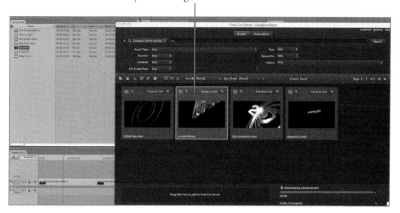

If no rough-cut edits were made, the bin will be populated with the individual high-res clips, ready to edit.

As you need more material from the server, you can press Command-Tab to access the Final Cut Server interface to search for more. If you have two displays for Final Cut Pro, you may want to leave the Final Cut Server window open on one of them. A reporter under deadline might be searching and finding more material to use, and then checking in additions to a project. Depending on how your notifications are set up, you could receive alerts as the reporter adds more material.

When you finish editing the story and are ready to go to air, you check your project back into Final Cut Server and change its status to Ready for Review (or Completed if there is no approval step). At this point, an email will be sent to the reviewer, who, on approving the project, changes the status to Completed. Then a new low-res proxy is generated, and the content will be marked as available for use in the newsroom.

This lets the news director and others in the newsroom confirm the story is ready to go. They can be in the building, or even remotely connected, in the case of the legal department needing to clear a sensitive story prior to going to air.

Should you need to connect remotely to a sister station or bureau with Final Cut Server, you can browse their assets and copy the high-res material locally on your system. This can add a level of efficiency to the production process that bypasses use of traditional satellite feeds for "element" gathering. Of course, available bandwidth governs how fast the transfer will be.

When a Final Cut project is checked in, Final Cut Server uploads any new media that may have been added to the project. This includes any new clips rendered by Final Cut Pro for review purposes. Notifications can be attached to this process. These progress bars show the generation of the proxies for the new clips in the project just checked in.

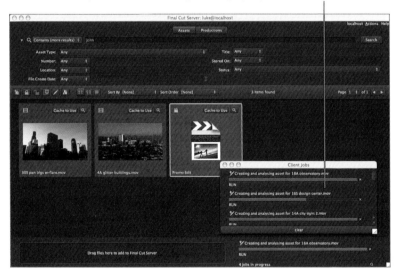

You've learned the primary methods of making available tapeless, tape, and now server material on Final Cut Pro for editing. Now it's time to edit a simple voiceover.

6
Editing Simple Voiceovers

Now that you have a basic understanding of how the Final Cut Pro user interface works and how you access material to edit, it is time to begin putting together a simple voiceover.

Contrary to its name, a *voiceover*, or VO, doesn't contain any recorded narration. A VO is a shot or sequence consisting of video and the natural sound recorded along with it, intended for the anchor or reporter to narrate, or "voice over," live from the studio or the field.

Voiceover material differs from, but is often used in conjunction with, content classified as *sound on tape*. In SOT segments, the recorded sound is intended to be in the foreground, without narration by on-air talent. An example of SOT might be a quote from an official or an athlete speaking to the camera.

News and sports sequences often combine VO and SOT material. A VOSOT segment begins with voiceover content (natural sound plus live narration) then segues into a sound on tape section, such as a celebrity quote. A SOTVO segment, by contrast, starts with SOT material, after which live talent picks up the narration for a VO section from the studio. A VOSOTVO piece sandwiches a "sound bite" between two pieces with live narration, and so on.

VO and SOT content are building blocks of more complex edited sequences called *package*s, self-contained sequences that include recorded narration and other elements. These are discussed in greater detail in Chapter 7, "Fast Package Editing," but this chapter covers fundamentals that will accelerate the process of editing full packages.

We'll begin with a discussion of several methods of delivering clips to the Timeline. You will no doubt gravitate to one or two you prefer, but you may find it useful to combine the various methods as well.

Storyboard Edit

A storyboard edit refers to a style of editing used in commercials and movies whereby the director and editor use drawings to create a rough visual outline of the commercial or scene. The director has an artist sketch the actors and scenery, showing camera angles for dialogue, prop placement, and the like, then uses the drawings as a framework not only for shooting but also later in editing.

Instead of sketching anything, we are going to use the clips we've shot as something akin to storyboards and organize them for a simple edit. Afterward, we'll refine the contents of those clips by trimming out what isn't needed.

1 Create a new project and sequence if you haven't already.

2 Save it to a location that is easy to find and give it a name that's easy to remember.

   This process activates the autosave function.

3 Control-click in the Browser and choose "View as Large Icons" from the shortcut menu. If you can't see all your clips in the window, you can adjust the size of the thumbnails by choosing "View as Medium Icons."

4 Position the clips you want from left to right and top to bottom in the order you want them in the sequence.

5 Lasso or Command-click the clips you want and drag them to the Timeline or the Canvas.

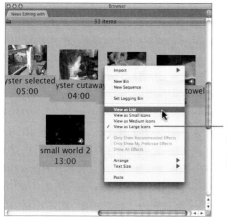

To view the clips as frames, Control-click in the Browser and choose "View as Large Icons" from the shortcut menu.

Be very careful when positioning clips in the Browser, as their vertical positions correspond to the order in which they appear in the sequence, with uppermost clips coming first. The clips should be arranged in something like a very slight downward stairstep.

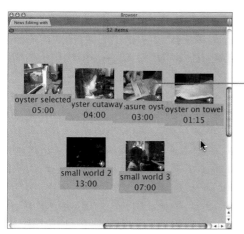

Arrange frames from left to right, each slightly lower than its predecessor. That way, when you lasso them and drag them to the Timeline, they will be sequenced as they appear here, from left to right, and top to bottom.

68 Editing Simple Voiceovers

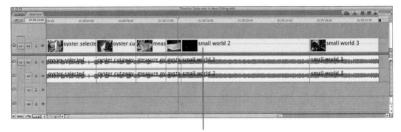

The clips then appear in the Timeline in the order in which their frames were arranged in the Browser.

When you drag clips from the Browser to the Canvas, the Edit Overlay appears in the Canvas, displaying seven colored sections representing different edit options. Drag the clips to the section representing the edit you want to make. If you're adding the first batch of clips to a new sequence, it doesn't matter if you drag them to Insert or Overwrite.

If you drag the clip to "Insert with Transition," a transition will be applied. FCP will automatically compensate and provide enough frames for the transition effect to be performed if none of your clips were marked with In and Out points.

Rearrange in the Timeline

Moving frames around in the Browser can be time consuming, especially in sequences containing many clips, and it may make more sense to drag the clips you want to the Timeline and then arrange them there, using a process called shuffle editing. With this method, you can just Command-click the shots you want in the Browser's Icon or List view, and drag them to the Timeline.

If you are using the List view in the Browser, you can sort alphabetically (which also organizes by the numeric suffix) or by timecode. You can lasso the clips you want or Command-click clips that are not adjacent, and drag them directly to the Timeline or the Canvas.

Linked Selection and Snapping

Before you begin arranging scenes in the Timeline, turn on linked selection (Shift-L) and snapping (N).

Linked selection limits Final Cut Pro's flexibility just a little, in a way that makes sense for news and sports editing: It ties together three clip components, or *clip items* (in this example, video, audio 1, and audio 2). FCP lets you use the default "on" to highlight one item; the other items will highlight automatically. Represented by parallel horizontal tracks in the FCP Timeline, video content appears in the track labeled *V*, and the audio components are found in tracks marked *A1* (audio 1) and *A2* (audio 2).

When linked selection is active, clicking a video or audio clip item selects the entire clip, not just the item in the clicked track. Because news and sports clips nearly always include live sound accompanying video, linked selection helps prevent sound and video from becoming separated during a hasty editing session. In addition to pressing Shift-L, you can toggle linked selection on and off by clicking a button, second from the right in the button bar in the upper-right corner of the Timeline.

The button doubles as an indicator as to whether linked selection is on; when the option is active, the icon on the button is tinted green.

The button to the right of the Linked Selection button is for snapping, which also should be turned on to streamline the process of rearranging clips in the Timeline.

When snapping is active, the head or tail of a clip you drag will jump to the playhead position when you get close, like a magnet. You can use this to precisely line up your insertion points.

Dragging and Dropping Clips

Moving shots around once they are in the Timeline is largely a matter of dragging and dropping, but be warned: 95 percent of the time in news and sports editing, you'll need to press the Option key between those steps—after you start dragging, but before you drop. If you just drag a clip and drop it, you will overwrite any content found at the drop location. Pressing Option before you release the mouse button turns on the insert edit mode, performing a Shuffle edit, which keeps content at the drop location intact and pushes it later in the Timeline, after the dropped clip. This is a very simple process, but also an important one you will use often.

> **NOTE** ▸ If you press the Option key before dragging a clip, you will copy the clip and move the duplicate. If you do this accidentally, press Command-Z to undo.

Select the Selection tool (press A) and click a clip (or Command-click a set of adjacent clips) to make a selection for rearrangement. When linked selection is active, clicking any component of a given clip will select all components.

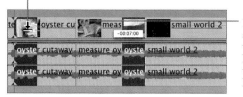

The down-arrow pointer indicates this clip, dragged without the use of modifier key, will overwrite any content at its destination in the Timeline.

Dragging this clip to a snapped-to transition point, without pressing a modifier key, causes the clip to overwrite the content at the destination location in the Timeline.

Overwrite edits leave gaps in the sequence, which is usually not desirable. If you create such a gap accidentally, press Command-Z to undo the edit.

To perform a proper Shuffle edit, follow these steps:

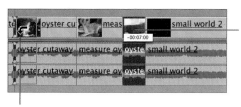

1 Drag a clip to your preferred location in the Timeline.

2 Just prior to releasing the clip, press the Option key. Note the pointer has changed into a curved arrow, or Shuffle edit pointer, which indicates an insert edit. Keep the Option key pressed until you release the mouse button. The clip will be inserted in the Timeline at the indicated location.

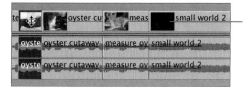

When you perform an insert edit with a new clip, the duration changes. If you drag clips around in the Timeline while pressing the Option key, you perform Shuffle edits, which do not change the duration.

If you drag a shot to the right to insert it, the Shuffle edit pointer changes direction.

The insertion happens the same way, bumping the shots to fill the space vacated by the shot you moved.

Using the Viewer with Clips from Videotape

You may have noticed that we have never suggested using the Viewer to apply In and Out points for fast editing.

It only rarely makes sense to do so when prepping a fast VO—when, for example, a long tape segment has been ingested in a single pass, yielding a giant clip that contains all the VO elements. Working that way contradicts the concept and practice of "fast editing" (wasting, say, 20 minutes loading a tape that will generate two minutes of finished content), so we won't detail the process of loading a clip in the Viewer, scrubbing through it to isolate shots, and gingerly marking each one's In and Out points, then adding each resultant segment to the Timeline.

It is far faster to throw the clips into the Timeline and trim them there. This has been proved by more than 15 years' experience in fast news editing on a variety of nonlinear editing systems by thousands of editors like you.

Using the Viewer with Tapeless Media

On occasion, you may get discs or cards that contain very long clips, such as press conferences or lengthy interviews. If you didn't organize the selected sound bites from the XDCAM or P2 interface, you can mark sections of the longer clip into smaller subclips in the Viewer. You should be general and less precise when doing this.

Because you loaded the material from the tapeless device in faster than real time, you haven't wasted much time (as you would have with videotape). Liberally mark In and Out points on a selected sound bite or element from the clip, and press Command-U. This makes a subclip, keeps the same clip name, and appends *Subclip* for the first instance, *Subclip 1* for the second, and so on. This lets you sort in the Browser and drag the correct bite when needed.

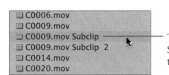

The first subclip comes in with no number. Subsequent subclips increment by 1 and keep the original clip name.

Topping and Tailing

To *top* a clip refers to starting it later than what was originally edited in the Timeline. To *tail* a clip is to end it earlier than what was originally edited in the Timeline.

Topping and tailing work well primarily while cutting voiceovers. The only time you would ever do this while cutting a package is to rapidly trim sound bites and standups in the Timeline. If you were to do a top and tail for B-roll in a package, you would have to be keenly aware of the relationship of the B-roll and the narration track. Therefore we will suggest the use of top and tail only for voiceovers and with sound bites and standups.

As yet these are not single-button functions in Final Cut Pro. There are two methods for doing this in Final Cut Pro: both are good, and both are faster than manually loading clips in the Viewer and carefully selecting In and Out points.

After you have arranged the shots in the Timeline in the order you want them, it is time to trim them to time. You may be reading a script of the anchor VO and trying to get the shots to match.

Keyboard Shortcuts for Topping

The following technique uses keyboard shortcuts to quickly top a clip by isolating and deleting the content preceding your new start point, and then scaling up the gap in the Timeline between the topped clip and the material that precedes it.

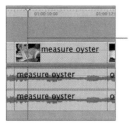

Park the playhead where you want the clip to start.

Press X to place an In point at the clip's first frame and an Out point at its last frame.

Press O to move the Out point marker from the last frame to the playhead location, thus selecting the portion of the clip you want to remove.

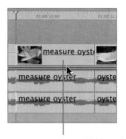

Press Shift-Delete to discard the marked segment and to pull the topped clip (and all unlocked tracks that follow it in the Timeline) back to close the gap between it and the clip that precedes it.

Keyboard Shortcuts for Tailing

A similar technique uses keyboard shortcuts to quickly tail a clip.

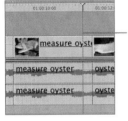 Park the playhead where you want the shot to end.

 Press X to mark an In point at the clip's first frame and an Out point at its last frame.

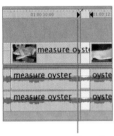 Press I to move the In point from the first frame to where the playhead is located, thus marking the area you want to remove.

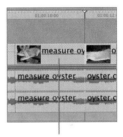 Press Shift-Delete to discard the marked segment and to pull back all unlocked tracks that follow the tailed clip in the Timeline, preventing creation of a gap in the sequence.

With just a little practice, you'll find that these keyboard shortcuts for applying tops and tails to clips becomes automatic and very fast.

While these methods rely on the video tracks for selecting new start and end frames for each clip, bear in mind that they also affect the audio tracks (assuming Auto Select is turned on). You can look for differences in the audio-channel waveforms for quick confirmation that a top or tail operation has been successful, but you should always play through your edits to make sure you don't introduce jarring audio transitions between clips. (We'll discuss correcting that problem later in this chapter.)

NOTE ▶ If a clip or segment elsewhere in the Timeline is highlighted while you are performing the top or tail "extraction" with Shift-Delete, you may notice that your edit didn't occur. Instead the highlighted clip somewhere else in the Timeline was removed. Undo the edit, click in the gray area above the Timeline, and repeat the Shift-Delete to top or tail the clip.

Using the Timeline and Viewer to Top and Tail

Once you have mastered the concept of topping and tailing, you may find it faster to sync the Canvas and the Viewer windows for fast-navigation trimming.

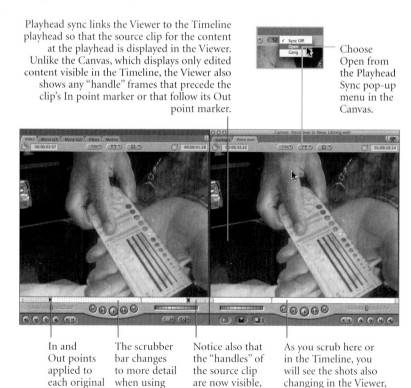

Playhead sync links the Viewer to the Timeline playhead so that the source clip for the content at the playhead is displayed in the Viewer. Unlike the Canvas, which displays only edited content visible in the Timeline, the Viewer also shows any "handle" frames that precede the clip's In point marker or that follow its Out point marker.

Choose Open from the Playhead Sync pop-up menu in the Canvas.

In and Out points applied to each original source clip.

The scrubber bar changes to more detail when using playhead sync.

Notice also that the "handles" of the source clip are now visible, but dimmed.

As you scrub here or in the Timeline, you will see the shots also changing in the Viewer, along with the existing In and Out points applied to each original source clip.

Topping and Tailing

 The Canvas doesn't provide frame-level clip navigation detail.

In point of clip

The Viewer reveals material preceding the clip's In point, which can be used to extend the clip so it starts earlier. (If there's material at the end of the clip, the Viewer reveals that as well.)

The Viewer with playhead sync turned on provides lots of navigation detail for the selected clip.

The white section of the Viewer scrubber bar indicates the section of the chosen clip that's visible in the FCP Timeline. Dimmed sections of the Viewer scrubber bar denote "handle" content for the clip, which is hidden in the FCP Timeline.

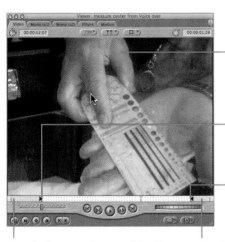

The Viewer now provides access to the entire contents of the selected source clip, not just the trimmed portion visible in the Timeline and Canvas.

In point for the clip, which marks the first frame of the clip visible in Timeline and Canvas views of the edited sequence.

Out point for the selected clip, which indicates the last frame of the clip visible in Timeline and Canvas views of the sequence.

Handle frames at the top of the clip, which you can use to extend the clip's Timeline duration by moving back its In point.

Handle content at the tail of the clip, which you can use to extend the clip's duration by moving its Out point.

Topping and tailing using the Viewer are very simple, and both use the same procedure:

1 Choose Open from the Playhead Sync pop-up menu in the Canvas.

2 Use the scrubber bar or the Viewer transport controls to locate the frame where you want your shot to start or stop.

3   Select the Ripple tool in the Tool palette or press the R key twice (RR).

4   Click in the Viewer.

5   Press I to set a new In point if you're topping the clip, or O to set an Out point if you're tailing it.

This updates the In or Out point in the sequence, shortens the selected clip, and seals up any resultant gap in the sequence to effect a ripple edit.

If your clip has hidden handle frames, you can use the same technique to extend a shot. In step 2 on the preceding page, drag the In point to the left or Out point to the right to extend. Clicking outside the bounds of the marked clip will take you to the adjacent clip in the Timeline.

Make sure the Viewer is active when you apply this technique. If the Timeline or Canvas is active when you press I or O, you will mark the In or Out points in the Timeline. If you do this by accident, undo the action (press Command-Z).

When you are finished trimming the clip, choose Off from the Canvas's Playhead Sync pop-up menu.

Conventional Trimming

All nonlinear editors have a trim mode. Final Cut Pro does as well, but removing material from the Timeline is much simpler with the top and tail methods. The conventional trim method is always an option and great if your deadline is next week, but it is time consuming. It is appropriate in fast-paced news editing only when you want to make a shot longer. But, since lengthening is the exception in news and sports, it's not covered in this guide. If you stick to one of the top and tail methods for shortening clips in the Timeline, you will finish much faster.

Now that we have the sequence laid out in the order we want it, and the time we want each clip to last, we need to concern ourselves with audio levels.

Simple Audio Adjustments

There are several ways to adjust audio levels. Included in Final Cut Pro 6 is an audio normalization option that sets the overall volume for your sequence (or a selected set of clips in the sequence) to the same level. Normalization raises the peak level to a selectable level, and then raises all the other levels by a relative amount. This is often handy for smoothing out differences in overall loudness between clips in voiceover sequences.

Before applying the normalization, mark In and Out points for the portion of the Timeline you want to normalize.

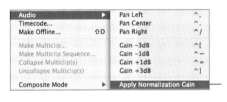

Choose Modify > Audio > Apply Normalization Gain. This adds a filter that makes the audio levels similar throughout the marked area.

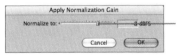

Adjust the slider or enter a numeric value to set the audio scaling for the marked area of the sequence.

If you need greater control over individual clip's sound levels, you can adjust their audio levels directly in the Timeline, or do it multiple ways in the Viewer. The Viewer is the easiest and largest display for audio adjustments.

Clipwide Level Adjustment in the Timeline

If you need to adjust a clip's overall audio level, as when its natural sound volume is noticeably louder or softer than that of adjacent clips, you can use the Clip Overlays control to make the change right in the Timeline. Select the clip in the Timeline and press Option-W to display clip overlays. Drag the audio level overlay up or down in the clip itself. To do this you need to click the Clip Overlays control at the bottom left of the Timeline, or press Option-W.

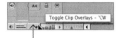

Click this button in the Timeline or press Option-W to display clip overlays, which causes a thin pink line to appear in all the audio tracks, indicating their sound levels.

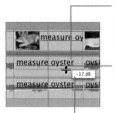

When you mouse over a clip's audio level overlay in the Timeline, the pointer changes to a Resize pointer. Drag the bar up to raise the clip's overall volume level, or drag down to lower its level.

FCP hides a clip's audio waveform during adjustment when its audio level overlay is displayed.

A tooltip displays the clip's dynamic level in decibels as you adjust the level up or down. When you release the mouse button, the audio level for the entire clip will be set to that level.

Ramping Clip Audio Levels in the Timeline

In addition to setting a clip's overall volume level, you can raise and lower the audio levels over select sections of a clip by manually entering keyframes. This allows you to automatically adjust the levels by varying the keyframe positions and levels, thus performing a mix not entirely unlike using a hardware mixer.

1 Select the clip you want to adjust by clicking it in the Timeline.

2 Click the Clip Overlays control (or press Option-W) to display the audio level overlay in the clip's audio track.

3 Press P to select the Pen tool.

4 Click the audio level overlay with the Pen tool to add audio keyframes, represented as pink diamonds, where you need them.

5   Drag an audio keyframe or the audio level overlay between keyframes up to increase the volume or down to reduce it. Drag keyframes sideways to change their position in the clip.

6   Audition the mix of the clip by backing up and pressing the spacebar.

Press P to select the Pen tool.

Click the audio level overlay with the Pen tool to add audio keyframes.

Click a keyframe or the audio level overlay between keyframes and drag up or down to ramp the audio level.

As you drag an audio keyframe (or the audio level overlay between keyframes), a tooltip displays the corresponding audio level setting in decibels. When you release the mouse button, the tooltip goes away.

To remove an audio keyframe, Option-click it with the Pen tool.

Another way to ramp audio levels from one portion of a clip to another is to cut the clip using the Razor Blade tool (press B), change the overall audio levels for one or both new clips, and then apply a cross fade to smoothly transition between levels. Before you split a clip, make sure snapping is active (press N if it's not on already), to prevent the creation of gaps between transitions.

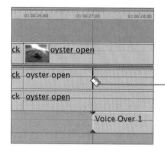

When you pass the Razor Blade tool over the playhead, it highlights the cut location. Click to split the clip at that spot. Deactivate the Blade tool to prevent inadvertent splitting of clips. Activate the Selection tool (A).

Drag the audio level overlay up or down to change the audio level.

Control-click (or right-click) the transition point between the clips and choose Add Transition 'Cross Fade (+3db)' from the shortcut menu.

The Cross Fade transition icon appears in the Timeline.

Clipwide Level Adjustment in the Viewer

Just as with video, editing audio in the Viewer gives you greater precision than you can get when adjusting sound in the Timeline or Canvas. The Viewer lets you see the audio waveform for a clip as you adjust its audio level. And in addition to letting you adjust the overall volume level over time, the Viewer lets you control *panning*, or the relative changes in volume between the left and right channels in a stereo sound mix.

1 In the Timeline, double-click the audio track of a clip you want to adjust.

   This opens the clip in a Viewer audio tab, with the playhead at the location you clicked in the Timeline.

2 Drag the pink audio level overlay up or down to raise or lower the audio level of that entire clip.

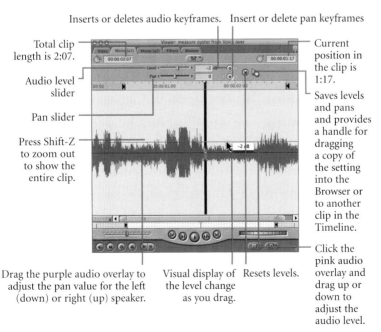

Ramping Levels Within a Clip in the Viewer

You can also ramp levels within the Viewer. Double-click the audio track you want to ramp. Select the Pen tool (P) and begin marking the points where you want to ramp the audio. Drag a point up or down (or earlier or later) to reflect the change.

Use the Pen tool (press P) to place audio keyframes at the desired locations in the audio channel.

You may also add keyframes to the pan overlay line (purple bar). Raising and lowering will adjust the pan settings for that track. This is a great way to have a mono sound move from the left speaker to the right speaker.

Drag keyframes or the audio level overlay between keyframes, up or down to change volume levels. Drag keyframes sideways to reposition them within the clip.

Audio Cross Fades

Audio cross fades are transition effects that prevent abrupt sound shifts between adjacent clips in a sequence. A cross fade essentially decreases the audio volume for the first clip while simultaneously increasing the volume for the second clip. The fastest way to perform an audio cross fade is to Control-click the audio transition edit point in the Timeline and choose Add Transition 'Cross Fade' from the shortcut menu.

Final Cut Pro offers two built-in audio transitions. The one you'll likely use most often is the +3 dB cross fade, which keeps the overall sound level constant as it shifts from one clip to another. The other option, the 0 dB cross fade, causes a dip in audio level at the edit point, an effect you may find desirable for certain situations.

The most basic (but not the fastest) way to apply an audio cross fade is to use Final Cut Pro's Effects menu. Click the edit point between adjacent clips' audio tracks, and then choose Effects > Audio Transitions > Cross Fade (+3dB).

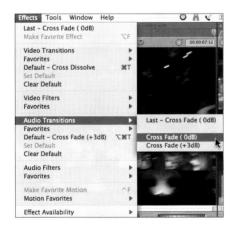

Final Cut Pro also designates a default audio cross fade, which is applied automatically when you click an audio edit point in the

Timeline and then press Option-Command-T. The +3 dB cross fade is the standard default. You can change it if you like (see below), but there's probably no reason to.

Applying Cross Fades in the Timeline

There are two ways to add an audio cross fade in the Timeline: by accessing a shortcut menu, or by using drag-and-drop.

For the shortcut menu method, Control-click (or right-click) the edit point between the clips you want to cross fade.

Your default audio cross fade will appear as a shortcut menu option when you Control-click the edit point between the clips' audio channels in the Timeline.

To change the default audio cross fade, click the Effects tab in the Browser and then click the disclosure triangle next to the Audio Transitions folder icon. Control-click the transition you prefer as the default and choose Set Default Transition from the shortcut menu. The same process also works for setting default video transitions.

To apply a cross fade using the drag-and-drop method, drag the icon for the desired cross fade from the Browser's Effects tab to the desired edit point in the Timeline.

> **TIP** Dragging an icon to apply an audio cross fade typically takes longer than the shortcut menu method, so it generally makes sense only for applying the cross fade that isn't your default.

Click the Effects tab

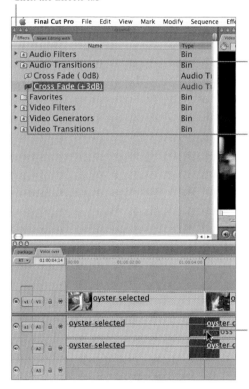

If necessary, click the disclosure triangle to reveal the contents of the Audio Transitions folder. The label for the default audio cross fade is underlined.

Drag the icon for the preferred cross fade to the desired edit point in the Timeline.

Release the mouse at the desired edit point to apply the cross fade.

Positioning a Cross Fade Between Clips

By default, cross fades center on the transitions between the clips they govern, lowering the volume level of the first clip and raising the volume of the second at the same time. You can have the cross fade start, center, or end on the transition, depending on available media.

Drag the cross fade icon in the Timeline. The pointer changes to the Roll tool, and a tooltip reveals how many minutes, seconds, and frames you are moving the cross fade off-center and in which direction. The number to the far right denotes frames.

A positive value in the Roll tooltip indicates a shift in the cross fade's center to a position later in the Timeline; a negative value reflects a shift to an earlier position.

Adding Dissolves

A *dissolve* is a video transition effect that has gained popularity in recent years with the advance of nonlinear editing systems such as Final Cut Pro. Perhaps because they can apply a dissolve with a just a few mouse clicks, rather than by jockeying three tape decks the old analog way, many video editors tend to overuse this effect. A dissolve is most effective when used to indicate a change in time, location, or both. Dissolves are more common in packages than in voiceovers.

There are a couple of ways to add a dissolve:

▶ Control-click at the desired edit point in the Timeline and choose Add Transition 'Cross Dissolve' from the shortcut menu.

▶ Click the desired edit point in the Timeline to highlight it, and then choose Effects > Video Transitions > Dissolve > Cross Dissolve.

If a Dissolve is your default video transition (it's the standard default), move the playhead to the desired transition in the Timeline and then press Command-T. FCP will apply a dissolve at that spot.

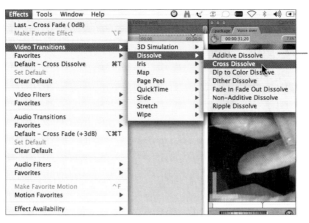

A wide variety of dissolves are available. The most common is the cross dissolve. Some entertainment shows use the "dip to color" dissolve frequently. Try them and see which you prefer.

Final Cut Pro's standard default dissolve transition, the cross dissolve, is basically the video equivalent of an audio cross fade. It fades the image in the first video clip to transparency, while simultaneously fading the second clip from transparency to full visibility. And like the audio cross fade, the position of a cross dissolve's "center" can be adjusted relative to clips it links. To adjust a cross dissolve's center, click its transition icon in the Timeline. This opens the dissolve in the Viewer's Transition Editor.

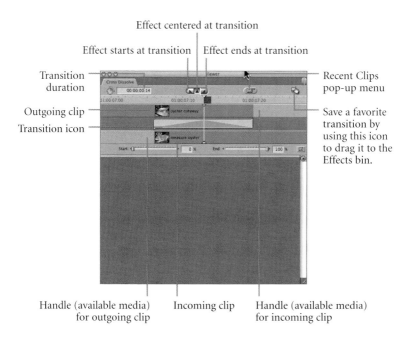

You can move the dissolve after placing it.
Drag the effect icon. The Roll tool appears.

The window shows the number of frames the transition has moved.

Outgoing clip, adjusted start frame

Removing a Dissolve

To remove a dissolve or other effect, select it in the Timeline and press Delete.

Fixing Material

Please see Chapter 8, "Quick Fixes," for more instructions on the use of video and audio effects for purposes such as the following:

- Adding a blur to disguise something
- Highlighting an item
- Resizing a clip
- Equalizing audio levels
- Fixing white balance

7
Fast Package Editing

A *package* is a self-contained, edited sequence consisting of a number of components. They include video clips with synchronous background natural sound (referred to as B-roll); sound bites (the portions of the interviews to be included in the story); a reporter's narration; and, perhaps, one or more recorded "standup" segments of the reporter at a location in the field. Also, at times, you may want to play some natural sound elements at full volume within the story. This is called opening for NATSOT ("natural sound on tape," a term left over from the analog era).

Packages are designed to play start to finish with no live in-studio contribution beyond an anchor's introduction. Therefore good storytelling requires that every element of a package—images, sound, and narration—be precisely positioned, synchronized, and balanced to help tell your story.

This chapter will describe a very fast method for putting packages together, one that may seem unconventional if you're used to "traditional" nonlinear digital editing techniques. This method has been developed, refined, and proven over a span of more than 15 years of nonlinear news editing experience on a variety of editing systems. This may not be the only way to edit packages quickly with Final Cut Pro, but it works very well.

Organizing for the Edit

We are assuming that you have already either captured the material you need from tape or absorbed media from a tapeless acquisition device, or that incoming feeds are available on your server. You will want to organize your material, sometimes using folders if you have time, one for B-roll, one for sound bites and standup takes, and maybe one for the narration track.

If you captured from tape, you may have already grabbed the sound bites leaving a few seconds of handles at the beginning and end of each clip to allow for backtiming or extending the clip later. This avoids having to search for a cutaway as the deadline approaches. Good, but don't spend time carefully marking the In and Out points on the tape for the bites. Remember, pre-roll is not our friend.

Typically, the narration is going to be available late and close to your deadline. Any preparation you can make prior to the delivery of the track, whether it's on another tape or available on the server is encouraged. You may also be planning to record the narration directly in the edit room.

Laying Out the Story

There are four steps to laying out a package, and they are not unlike the way we old-timers used to edit news stories with film. Some of us even edited videotape like this, but because tape is linear, we didn't have the advantages of, for instance, laying in the last clip first and working backward (not that you would do that every day).

The four steps are very clear, and the screen shots on the following pages show the progress of the story:

- ▶ Full Track—The entire recording of the track is put in the Timeline, bad takes and all.
- ▶ Tight Track—The bad takes are deleted, and a seamless "script read" is produced.
- ▶ Fat Bites—Unmarked sound bites and the standup(s) are inserted among the narration elements.

► Tight Bites—The sound bites and standups are trimmed. This could also be called the *radio cut*. If you shut off the picture, you should hear, seamlessly, the components of the script.

At Tight Bites, you begin laying in B-roll and opening for nat sound. After that, you adjust your audio levels, add any effects, and output either to tape or export to your output server.

Recording the Narration

Select the Voice Over tool from the Tool palette, or press Option-0.

The Voice Over tool lets you record directly into the Final Cut Pro Timeline, with the media being recorded locally on internal drives or directly onto the server. Press Option-0 or choose Tools > Voice Over to use the tool.

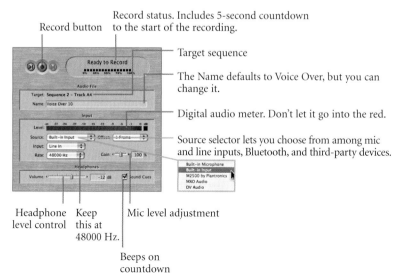

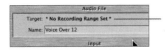 If you see "No Recording Range Set" as the target, you need to preset a Timeline duration to continue.

 To set the duration, click in the left Timecode Duration field in the Canvas. Type *50000* (5 minutes) and press Return.

If you already have video in the Timeline, it will drop the recording on the next available audio track. If you have a new Timeline, you will need to set a duration in order to use the Voice Over tool. It may be easier to just throw a clip into the Timeline to avoid having to type an arbitrary duration.

The Voice Over tool will default to audio track 2 if you have no source clip in the Viewer. Because we will be using tracks 1 and 2 from the source material, you can easily drag the entire recording to audio track 3 prior to deleting the bad takes.

You can also set the recording to go straight to the A3 destination track in the Timeline:

Controls in this column represent the Source video and audio channels associated with the clip in the Viewer. Source channels are labeled using lowercase letters. This clip has a video source track, v1, and audio source tracks a1 and a2.

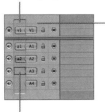 This column contains Destination controls for the project's video and audio tracks, as shown in the Timeline and Canvas. Destination channels are labeled using capital letters. In this example, the v1 source channel for the clip is "patched" to destination channel V1; source channels a1 and a2 are patched to destination channels A1 and A2; and unused destination channels A3 and A4 are available for recording narration.

To record directly onto track A3, load a clip with two audio tracks into the Viewer. Then drag the a2 Source control and snap it to the A3 Destination control. The Voice Over tool will display the new destination setting.

If you want to save and rename this recording, it is a good idea to grab it in the Timeline and drag it to the bin. As a matter of having to find it later for a second version of the story, you should rename the various narration elements.

Total duration of this recording is 1:27 (minute:seconds). This includes the bad takes as well as the good.

Another caveat is that while recording the track you can click other parts of Final Cut Pro, but if you double-click anything or switch to another application, you will stop the recording. Because of this, and because alert sounds from background programs can ruin a recording, it is a good idea to close all applications other than FCP while recording narration or capturing video from tape.

Depending on how you choose to monitor the reporter's progress, it is suggested that you take notes on the good takes.

It is a good idea to read along with the script and mark the take numbers so you can skip to the right take when cleaning up. The reporter should be slating the takes, followed by a three-two-one verbal countdown leading into the track, so you can find them more easily (for instance, "track 2 take 4"). This way, once you finish cleaning up track 1, you can skip through the first three takes of track 2.

Cleaning Up the Full Track

There is no method of marking the narration while it is being recorded to flag the good takes, so cleaning up the track cannot be done automatically. The manual procedure is still pretty fast, however.

By activating the waveforms in the track, with a little imagination you can see where the "three-two-ones" are located for easy navigation.

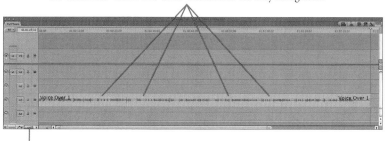

Choose Show Audio Waveforms from the Track Layout pop-up or press Option-Command-W.

1 Cue the track to the beginning.

2 Play or scrub down to where the first good take begins and stop.

3 Using the topping keyboard shortcut sequence (press X, O, and Shift-Delete), lop off the left side of the track in the Timeline.

This will pull up the sequence so that the first frames are right at the first piece of narration.

You should always listen to the narration at the transition points to ensure you have the right takes in the story.

If your reporter is leaving a gap of a second or two between takes, slating the individual takes, as well as providing a "three-two-one" verbal countdown, it will be much easier to identify the individual takes using the audio waveforms. By making audio waveforms visible in the Timeline, you can quickly see where the good and bad takes are by counting the gaps in the waveforms. You may want to scrub past the beginning to confirm a take, but you certainly don't want to play through all of the takes.

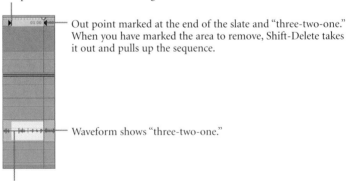

In point marked at the end of a good take.

Out point marked at the end of the slate and "three-two-one." When you have marked the area to remove, Shift-Delete takes it out and pulls up the sequence.

Waveform shows "three-two-one."

Large gap shows break in track.

This yields what is called a Tight Track, which runs 0:42.

Five track regions, with four edits

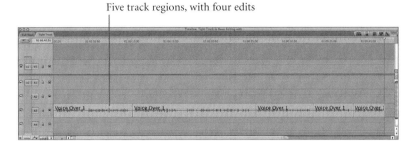

Another way to clean up narration tracks is to use the Razor Blade tool (press B) to divide the track into the good and bad takes.

The Razor Blade tool breaks the narration into two clips. This position is just at the end of a good take leading into the "three-two-one" countdown on the slate for a bad take.

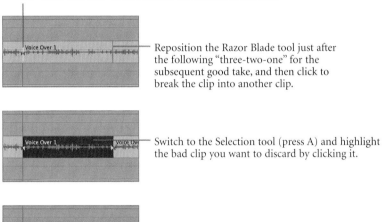

Reposition the Razor Blade tool just after the following "three-two-one" for the subsequent good take, and then click to break the clip into another clip.

Switch to the Selection tool (press A) and highlight the bad clip you want to discard by clicking it.

Press Shift-Delete to discard the bad clip which pulls up the remainder of the narration track to "seal" the region that the discarded clip vacated. It is always a good idea to play across the transition to ensure the proper content.

Inserting Sound Bites and Standup(s)

Assuming you already have your sound bites and standups isolated in the bin, we will now be throwing them in the Timeline, but to specific places.

 You will want to ensure that snapping is on in the Timeline. The N key toggles it on and off.

One method of organizing sound bites, standups, and other material from longer recordings is to make "subclips" that isolate the elements into individual short clips so they are ready to use. Perhaps you have an hour-long feed that came in, which displays as one clip. You can pull certain elements from that one hour by subclipping, making it easier when you have to put the story together.

When you mark In and Out points on a clip in the Viewer and then drag it to the bin, the entire source clip, not just the visibly marked portion, is added to the Bin. Multiple edits from the same source clip, therefore, lead to multiple copies of that clip in your bin, which can needlessly bloat the size of your project.

You can avoid this by breaking your long master clip into shorter segments, or subclips, before you mark your final In and Out points. To define a subclip, mark an In and Out in the Viewer, and press Command-U to create the subclip and deposit it in your logging bin.

When creating subclips, focus on breaking up the longer clips into the number of shorter subclips you need, but don't spend a lot of time locating their exact In and Out points. You can do that more easily in the Timeline.

If you're working with Final Cut Server, a producer or writer could be doing this as the feed is being recorded, so that you get material that is already subclipped.

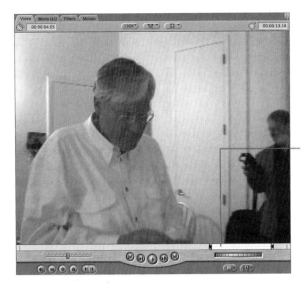

Load the clip in the Viewer, find one of the sound bites, and click the Mark In and Mark Out buttons. Choose Modify > Make Subclip (Command-U) to deposit the marked area in your bin.

Highlight the name of the subclip in the bin and rename it *1st bite*, or *Bite 1*, or whatever you wish. Continue through the larger clip extracting the general regions of subsequent bites and renaming them so you can find them easily.

TIP Adding the first two or three words of each sound bite to its clip name is a handy way to identify each clip and ensure its correct placement when you're under deadline pressure.

Dragging Bites to the Timeline

One of the many nice things about Final Cut Pro is that you don't have to have the playhead sitting on the transition point where you want the edit to be spliced. You will use snapping to choose where you want the bite to be inserted between narration elements. This can all be accomplished without use of a modifier key. You just need to be careful when and where you drop the clip. Before you make the drag and drop, identify in the video track a thin gray line about two-thirds up from the bottom of that track. When you drag the clip, let it go above that line, when you see the insert icon (a right-pointing arrow).

The wrong way:

This is the 2/3rd line in each of the tracks. Placement above or below this line determines the type of edit, Overwrite is below the line, insert is above.

The resulting edit replaces the narration. If this happens, undo.

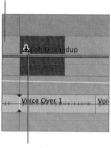

Make sure snapping is on and then drag the standup to the desired transition point. The downward-facing arrow indicates that this will be an overwrite edit, not the edit you want.

The correct way:

Instead, drop the clip slightly higher in the Timeline, when you see the arrow change from pointing down to pointing to the right.

This inserts the standup into the sequence and pushes all tracks down, lengthening the sequence.

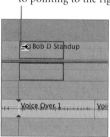

When you insert the shot, the Timeline expands and may push the rest of the sequence elements out of view. Press Shift-Z to bring the whole composition back into view.

It is always a good idea to play the last second or two of a track and to check against the script to confirm you are placing the bites in the proper location.

Continue throughout the sequence dropping in the subsequent bites and standups in relation to the narration tracks.

Once you have completed splicing these elements into the Timeline, you are at the Fat Bites stage. This package still needs to be trimmed down to the actual sound bites and standup durations.

Because every source clip contains a video track and two audio tracks, inserting the voiceover clips into the narration track creates a series of gaps, three channels "deep," that need to be filled with B-roll to complete the package.

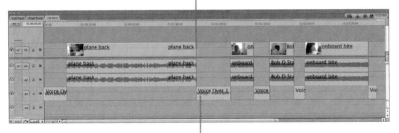

We have put the narration on track A3 so we can take advantage of the clip boundaries.

Trimming to Tight Bites (the Radio Cut)

We now bring back the top and tail methods from Chapter 6. These lop off the first and last few frames or seconds of the sound bite and standup components of the story. As you will see, doing this is much faster in the Timeline than ever-so-carefully searching for an In point in the Viewer.

> **TIP** The top and tail methods require that the Auto Select controls on the Timeline are active. By default, they're active, but if they are turned off, topping and tailing may result in losing sync farther down in the story.

First, the top operation:

1. Cue just before the first sound bite.
2. Click the Play button.
3. Press the spacebar to stop exactly at the point where you now want the sound bite to start.
4. Scrub the clip frame by frame in case you missed the In point.
5. Press X, O, then Shift-Delete.

Elements of the sound bite are highlighted and removed, and the sequence is pulled up. This is called a *ripple* edit, as it changes the duration of the sequence. Now, the tail operation:

1. Cue to just before the end of the sound bite.
2. Click Play.
3. Stop exactly at the end of the bite, if you miss it, scrub frame by frame either way to get to it.
4. Press X, I, then Shift-Delete.

The end of the bite is marked and removed, and the sequence is pulled up, or "rippled."

Continue through all the sound bites and standups. Once the operation is complete, you have the radio cut and are at the Tight Bites stage. You can give your total time to the producer, or you can add some natural sound (as described later in this chapter) and then give the show producer the total run time. At the Fat Bites sequence, the TRT was 2:32. Now at Tight Bites it's at 1:22.

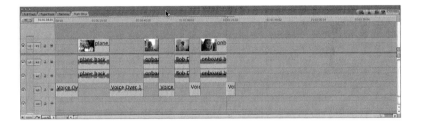

Adding B-roll

We now have an edited story, comprising narration and sound bites, with a runtime of 1:22. Now we need to add video to accompany the narration. This is easily accomplished with a very slight modification of the top and tail methods.

1. Play the sequence from the beginning and stop where you want the first clip to end.

2. Mark the clip by pressing X, and update the Out point by pressing O.

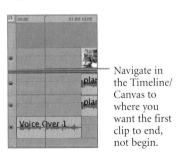

Navigate in the Timeline/Canvas to where you want the first clip to end, not begin.

Press X to mark the In and Out points for the vacant area, so you don't have to go to the first frame and mark it manually.

Press O to move the Out point to the playhead, thus isolating the area where you want the edit to be made.

Drag the clip to the Canvas and drop it when you see the down arrow in the Overwrite area. The clip conforms to the marked area in the Timeline.

Unfortunately, you can't drop it in the Timeline as we did the sound bites. The entire clip would cover your sound bite, even though you have marked an Out point in the Timeline. Your options are using the F10 key, clicking the Overwrite button, or dragging to the Overwrite section of the Canvas Edit Overlay.

Drag the clip to the Overwrite section of the Canvas Edit Overlay. The clip conforms to the marked area in the Timeline.

Or click the Overwrite button (F10).

Be aware that if you don't mark an In point, the clip will be edited in from the first frame. You can still "slip" the edit, as described later, but you may want to mark an In point in the Viewer to get you in the ballpark.

Adding Pad at the Top

Since this is the first shot in the story, you most likely will need to have a second of pad at the top before the narration begins. Videotape required at least a second to come up to speed before it "locked." Servers are much faster, but human reaction times are not, so the custom persists.

Pad helps prevent the director from punching up "black" or a still frame of your story on the air. It allows the control room to roll your story and have the director "take" once video is playing. If your narration starts without any pad, the first word or two could be clipped.

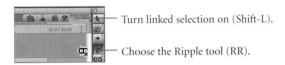

— Turn linked selection on (Shift-L).

— Choose the Ripple tool (RR).

 Clicking the start of the sequence should light up all three track transitions and exclude the narration since it isn't linked.

 Click one of the transitions and drag to the left, watching the counter. When it is around −1:00, stop and release the mouse button.

 The entire sequence has been pushed earlier in the Timeline to reflect the time that has been added as a pad.

Alternate First Shot Method

You will always need about one second of pad at the top of a story, if only to compensate for the possiblilty of slow reaction time in the control room. In some situations, you also may want to do what is called a "cold open." This is where pertinent natural sound is played at full volume prior to the start of the reporter's narration.

1 Find the clip you want to use and load it in the Viewer.

2 Mark an Out point at the end of the area where you want the sound to be up.

3 Navigate to 1 second prior to the actual sound you want to start the story, and mark an In point.

4 Insert that edit into the Timeline, bumping the rest of the sequence to the right, including the narration.

At this point, you will want to extend the clip over the narration and fade the audio. To do so, play the Timeline to where you want the first clip to end, and pause.

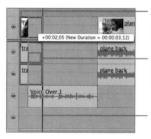

You have spliced the sound-up cold open with only the duration of the sound up. This bumps the rest of the package to the right in the Timeline.

Play to where you want this first clip to end. Click the transition point and, with snapping on (N), drag it to the playhead. The shot snaps to it.

Extend the incoming shot to the playhead, or drag the transition point to the playhead.

Use the Pen tool (P) to put keyframes on the audio level overlay. Drag the last two to a lower level to compensate for the narration track.

This works extremely well and is a fast alternate method for placing the first clip in the package.

> **TIP** Once you fill a B-roll area up to a sound bite or standup, you should adjust your audio levels using the techniques described in Chapter 6, "Editing Simple Voiceovers." Consider using normalization as a first step toward adjusting your natural sound levels. Mixing as you go will prevent getting all the way through your story covering the B-roll and not having enough time to adjust audio levels. If you work on audio levels sequentially as you add each sound bite, you could very easily make a judgment to drop in a 17-second single shot, for instance, as the closing shot as your deadline approaches. Better to have a proper mix for the whole story and a long closing shot than a bad mix with a variety of shots at the end.

Adding More B-roll

Continue filling the B-roll areas by pressing X to mark the clip, then O, and then dragging the clip to the Overwrite section in the Canvas Edit Overlay (or pressing F10).

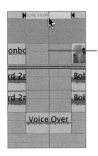
When you are coming up on a sound bite and want to insert a single shot to fill the gap, you need only move the playhead into that gap and then press X to mark the In and Out points for that area. You can do this while the sequence is playing, or you can drag through it, snap to transition points, or click in the area you want to mark.

Slipping

A slip edit shifts all the frames of a clip forward or backward, without changing the clip's duration or its position in the sequence. A slip might, for instance, move a clip's start ahead 12 frames and also shift its end frame (and every frame in-between) ahead 12 frames as well. Slip edits obviously require that there be enough handle frames on each end of the clip to accommodate any shift. It won't slip any farther than the frames available. Slipping is a great way to tweak the content in the Timeline without changing the sequence's total duration.

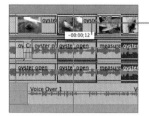

To perform a slip edit, highlight the clip, making sure linked selection is on, so you need only click one component of the clip. Press S to select the Slip tool, and then drag the clip to the left to start the clip on an earlier start frame, or to the right to shift to a later start frame. The counter updates as you drag the frames backward (negative numbers) or forward (positive numbers).

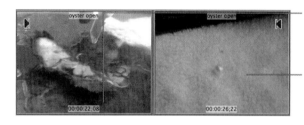

As you drag in the Timeline, the first frame of the clip updates.

The last frame of the clip also updates while you drag in the Timeline.

Sliding

A slide edit shifts the position of a particular clip in relation to the clips that precede and follow it, without changing its duration or the total running time of the sequence. A slide entails picking up the clip and dragging it forward or backward in the Timeline. If, for instance, you slide a clip 17 frames backward in the Timeline, 17 frames are trimmed from the end of the clip leading into it, and 17 frames are added to the start of the clip that follows it. Sliding is a great way to reposition a shot at a specific point in narration or music.

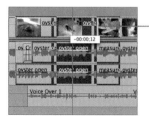

To apply a slide edit, click a clip (with linked selection on). Press SS to select the Slide tool and drag to the right or left. A bold outline will appear around the clip to highlight the movement, and a pop-up counter will display the time and number of frames the clip is shifted. Negative numbers indicate movement toward the start of the sequence; positive numbers indicate movement toward the end of the sequence.

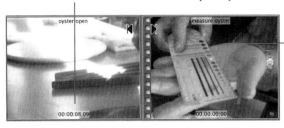

As you drag a clip to the left, the display updates to show the last frame of the clip that precedes it.

The first frame of the clip that follows the one you're sliding appears here while you are dragging in the Timeline.

Backtiming

As a way of introducing a sound bite subject, or to compensate when a B-roll clip comes up short against your narration, leaving a gap in the video and natural-audio tracks, it is sometimes useful to extend, or *backtime,* your sequence by a few seconds. This adds frames to the sound bite that fills the gap, so that the sound bite audio fades up and overlaps the narration track for a few seconds.

Make sure linked selection is on and press A to select the Selection tool. Drag the playhead to the spot where you want the video to start. In this case, that's the end of the B-roll clip, preceding the gap in the Timeline.

Click the clip you want to backtime, in this case the one following the gap, and then press E to extend the clip back to the playhead.

Once you've backtimed a sound bite clip so that it overlaps the narration track, you will want to adjust the sound bite clip audio so its level fades up, instead of starting at full volume as the narration ends:

1   Zoom into the track by pressing Command-+ (plus sign).

2   Click the Clip Overlays button to display the overlays, press P to select the Pen tool, and click the audio level overlay to insert two audio keyframes just before the bite starts.

3   Drag the left keyframe down a little. If you have audio on both tracks, see which is the dominant and adjust only that one.

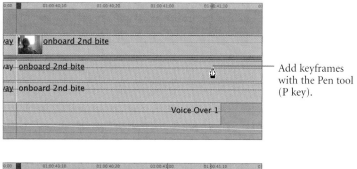

Add keyframes with the Pen tool (P key).

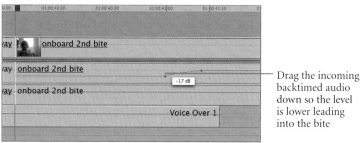

Drag the incoming backtimed audio down so the level is lower leading into the bite

Another way to fade the audio is to use the Razor Blade tool to break the shot in two pieces. This is described in Chapter 6, "Editing Simple Voiceovers."

Padding the End of a Sequence

There's always a chance the control room will switch away from your story a little late. To prevent the story from going to black if that happens, it's a good idea to leave a video "pad" that extends the sequence a few seconds beyond the end of your narration track.

Move the playhead into the gap where you want to insert the B-roll for the closing clip and press X to mark the remaining duration of the narration track. Open the clip you want to use as your final shot

Padding the End of a Sequence 111

in the Viewer and mark only an Out point, selecting a moment that provides a solid "payoff" to conclude the story. Press F10 to overwrite the edit, and FCP will do the math to find the correct In point to fit the clip into the Timeline gap. Play the clip and see if you like it, and tweak the In and Out or slip the shot as needed.

Once the final clip is fit for time to the narration, we need to extend it a little beyond the narration, to provide a pad that prevents bad video (or no video) from getting to air, in case the control room switches away late.

The easiest method is to extend the edit into black. To do this with the Selection tool active (press A), click the transition point (with linked selection on) to select tracks V1, A1 and A2. Click at a point a couple of seconds later in the Timeline, and then press E to extend the clip.

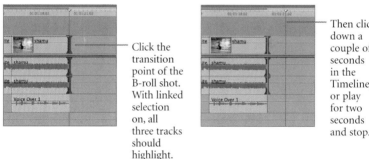

Click the transition point of the B-roll shot. With linked selection on, all three tracks should highlight.

Then click down a couple of seconds in the Timeline, or play for two seconds and stop.

Press E to extend the selection to the playhead position.

If you'd like, you can use the Pen tool (press P) to add audio keyframes to lower the audio level at the end of the sequence.

Another method is to press R to select the Roll tool, and then click the edit point at the end of the final clip and drag it to the right.

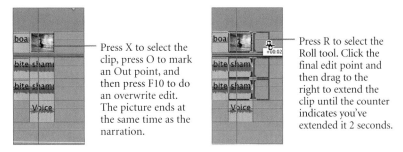

Press X to select the clip, press O to mark an Out point, and then press F10 to do an overwrite edit. The picture ends at the same time as the narration.

Press R to select the Roll tool. Click the final edit point and then drag to the right to extend the clip until the counter indicates you've extended it 2 seconds.

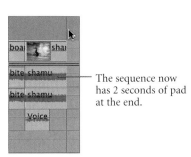

The sequence now has 2 seconds of pad at the end.

Opening for NATSOT

Opening for NATSOT, or inserting natural sound on tape, is an editing technique that creates a gap in the narration track to highlight a natural sound. It is a technique that gives your story breath and pacing. It can either be done as you edit, after you finish a section, or at the end of the editing process.

There are a couple of ways to add natural sound, and both are based on ripple edits.

You can open for NATSOT anywhere in the story. In this case, we've navigated to the edit point between a sound bite and a B-roll clip of someone tapping an oyster shell five times to crack it open.

Here, we use the Roll tool to extend the start of B-roll clip, inserting the NATSOT open audio and video (the first three taps) between the sound bite clip and the start of the narration:

1   Navigate to the start of the narration you want to open for NATSOT.

    In this example, that's at the end of the sound bite and the start of B-roll. We want to extend the opening of the B-roll clip and use its natural sound as a transition into the narration.

2   Press RR to select the Ripple tool. Click the edit point between the sound bite clip and the B-roll to highlight it. Press U to rotate through selection of the Out point of the first clip (which highlights the left side of the edit point, as shown), the In point of the second clip (highlights the right side of the edit point), and both. We're going to ripple back the In point of the second clip, so make sure the right side of the edit point is highlighted. Also make sure the narration track in channel A3 is not highlighted.

3 Drag to the left to extend the shot.

   A counter appears to show you how many seconds and frames you are adding to the start of the clip. Our example opens the sequence by 1 second and 23 frames, and pushes all the content in the sequence forward, making the entire sequence 1 second and 23 frames longer.

The result of the NATSOT open.

The sound of the NATSOT open will play at full volume, then we'll lower the natural-sound level as the narration begins:

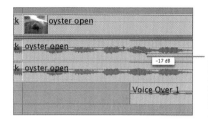

At the end of the waveform for the sound burst you want to play at full volume, use the Pen tool (press P) to insert two audio keyframes on the audio level overlay. Drag the second one downward to lower the natural sound level where it overlaps the narration track.

Natural Sound "Pops"

Allowing natural sound to "pop out" of the background between sentences in the narration can add immediacy and atmosphere to a story. Rather than trying to synchronize the reporter's narration around a desirable NATSOT accent, the following method lets you drop a quick natural-audio burst or "sliver" into the narration, making it sound as if the sound were timed perfectly to punctuate the story. These audio exclamation points can enhance your story, but be careful not to overuse the technique.

Find a suitable location in the narration for a quick sound burst, which could be a door slam, the crack of a baseball bat, someone on camera saying something brief, or, in this case, someone tapping an oyster shell five times to crack it open.

The following steps describe how to "open up" the narration track at an opportune spot for a sound sliver; how to insert a B-roll clip containing a sliver, so that its sound comes at the right moment in the narration; and then how to make sure the narration track and the NATSOT pop blend together seamlessly.

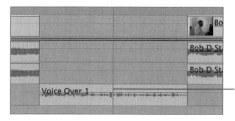

Navigate to a natural break in the narration track.

116   Fast Package Editing

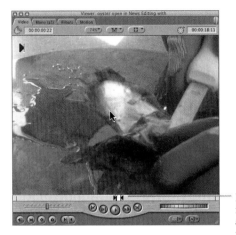

In the Viewer, open a source clip from which you want to grab the audio burst. Mark In and Out points around the sound burst you want to insert into the narration.

Click the Viewer's Mono (A1) tab to see audio detail.

The marked area includes the sound. Drag the In and Out markers as necessary to isolate the sound's waveform.

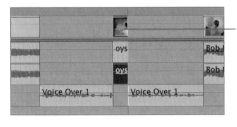

Insert (F9) the marked area into the Timeline to open the entire sequence across all tracks for the duration of the marked area. The sequence becomes longer by that amount, in this case 22 frames.

Now, to make the sound accent seem natural, you will want to back-time and extend it, so it overlaps and blends with the narration:

1 With linked selection on, press A to select the Selection tool.

2 Click the left edge of the clip containing the newly added sliver.

3 Drag the playhead to where you want the shot to start in the Timeline. If you want the shot to start at the end of the clip preceding it, it helps to have snapping active (press N).

4 Press E (extend).

5 Select the end frame of the sliver clip by clicking it, or by pressing the Down Arrow key.

6 Drag the playhead to the spot where you want the shot containing the sliver to end.

7 Press E (Extend).

As long as there are enough frames in the handles preceding and following the sound burst, the inserted clip expands to fill the gaps preceding and following it.

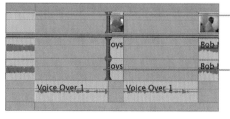

1 Drag the playhead to the end of the shot that precedes the clip containing the sound sliver you're inserting into the narration.

2 Press A to select the Selection tool, and, with linked selection on, click the clip containing the sliver to select all its tracks.

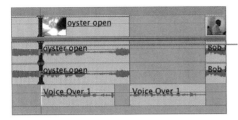

Press E to extend the shot back to the playhead.

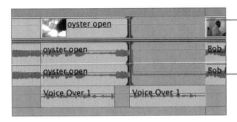

1 Drag the playhead to the beginning of the shot that follows the clip containing the sound sliver.

2 If it's not still highlighted from the preceding step, press A for the Selection tool and click the end frame of the "sliver" clip.

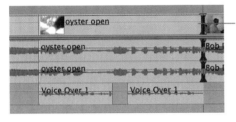

Press E to extend the shot forward to the playhead.

All you need to do now is adjust the B-roll sound level so it rises to full volume as the sliver begins and then returns to background level after the sliver.

1 If clip overlays aren't already visible, press Option-W to display them.

2 Press P to activate the Pen tool.

3 Click the audio level overlay to insert a pair of audio keyframes on either side of the pop in the sliver clip (four keyframes total).

4   Drag down the opening keyframe to lower the audio level behind the narration and make it ramp up going into the pop, and then drag down the closing keyframe so the level drops again following the pop.

When you play it back, the sound pop should sound natural within the edit.

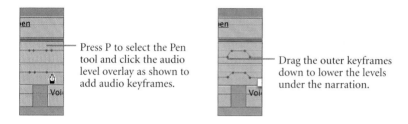

Press P to select the Pen tool and click the audio level overlay as shown to add audio keyframes.

Drag the outer keyframes down to lower the levels under the narration.

Video Effects and Rendering

The same methods apply for packages as for voiceovers. A full description of adding dissolves is in Chapter 6, "Editing Simple Voiceovers." Other effects are covered in Chapter 8, "Basic Fixes." Information on rendering and delivering the story is found in Chapter 9, "Delivering the Story."

Finishing

The final steps in completing a package include cleaning up all audio and video tracks, including dropping unwanted audio tracks and adding transitions between clips.

To clear out unnecessary audio, turn off linked selection and then Command-click the audio tracks you want to discard, and then press Delete. This removes the unnecessary track(s) and does not affect the total story time.

1 Make sure linked selection is off.

2 Identify any extraneous audio tracks and Command-click them to select them.

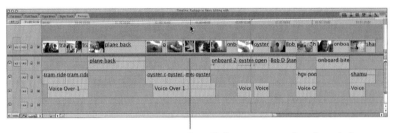

Press Delete to remove the selected clips.

It is always a good idea to play back the story as you work on it to check for content and pacing. Once you are satisfied, it is time to deliver the story for playback.

Because playback methods vary from organization to organization, playback devices have different requirements. Delivery details are best covered by on-site training at your workplace.

8
Basic Fixes

You will find the need to apply certain types of effects all the time, not necessarily to add an artistic component, but to fix something. Perhaps the material was shot with the wrong filter, or you have to blur someone's face in a crime story. Several types of effects are described here, with ways to simplify their use.

Motion Effects

At times you may need to stretch limited material to fit into a predetermined space in the story. In other cases you may want to run a sports play in slo-mo. Both of these are motion effects and can be achieved in different ways.

Fit to Fill

If you have file material that lasts 4:02, as in this case, and you have a space in your story that is 6:09, you need to do what is called a fit to fill edit. Final Cut Pro does the math to determine how much to stretch the video by slowing it down to fit that space.

Fit to fill can also be used to accelerate video. For instance, you may be trying to edit a very long pan or tilt shot into a short space on the Timeline.

If you follow the steps on the following page, as long as there are four points marked (In and Out on the original clip, In and Out in the sequence) and you drag the clip to the Fit to Fill section of the Canvas Edit Overlay, the math will be done automatically.

You needn't dwell on the duration of the clip you're inserting, since a fit to fill edit will make the clip fit the allotted space in the Timeline.

Having marked the target area of the Timeline or Canvas, the duration appears here. You can mark a random blank section of the Timeline as well, say 10 seconds. The calculation of the Fit to Fill would then take the source clip duration and make it fit into 10 seconds.

Mark In and Out points in the Viewer.

Drag the marked clip from the Viewer to the Fit to Fill section of the Canvas Edit Overlay. The clip will be sped up or slowed down as needed to fit the selected region of the Timeline.

NOTE ▶ If a blank region marked in the Timeline includes audio channel(s), any audio from a clip inserted via fit to fill is speed-shifted and kept in sync with the video. That means, for instance, that the crack of a starter's pistol at a track meet will slow down to sync with a stretched clip of a track meet. With any editing tool other than Final Cut Pro, keeping audio and video synced for this type of edit is at best cumbersome and often impossible.

As with all its motion effects, Final Cut Pro gives you full control over fit to fill edits and the precise manner in which the action (and sound) in a clip are squeezed or stretched to fill a span within the Timeline. By default, FCP speed-shifts each clip equally, and that's fine for many situations, but you can tweak the shifts as desired. You could, for instance, make a clip ease gradually from normal speed to slow motion (or hyperspeed). Controls for this are found in the Viewer's Motion tab.

Scroll down to Time Remap and click the disclosure triangle if necessary to reveal settings options.

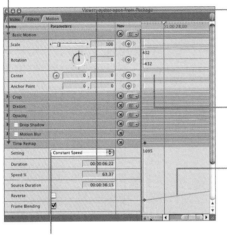

The Speed % indicator shows the extent of the fit to fill speed shift as a percentage of the clip's original speed. Values less than 100 indicate the clip has been slowed down, and those greater than 100 indicate it's been sped up.

Double-click the clip that you edited in the Timeline to open it in the Viewer, and click the Motion tab.

The Setting pop-up indicates that the default setting, Constant Speed, has been applied to this clip. The only other option is Variable Speed.

The time graph charts the rate of speed change for the clip. A straight line indicates constant speed, and a 45-degree incline does not necessarily mean 100% speed and depends on the granularity (zoom) of the Timeline display. If you click the graph with the Pen tool (press P), you can add motion keyframes and drag them to vary the rate of speed change within the clip.

Using the Motion Tab

Final Cut Pro lets you apply speed adjustments to any clip in the Timeline, not just those inserted via fit to fill. Motion can also be applied to the original clip, but stay away from that as it is a permanent change to that clip. To apply a motion effect for a clip already in the Timeline, double-click the clip in the Timeline to open it in the Viewer, and then click the Motion tab.

Use the same Time Remap controls to adjust fit to fill edits and to vary playback speed within a clip. These controls have many variables, and they can be confusing under deadline conditions, so experiment only when there's time to spare.

Type here to change the speed setting applied to the clip. Larger values increase clip speed; negative values run it in reverse. If fit to fill sets this value automatically, changing that value here will increase or decrease the clip duration and apply a ripple edit that changes the overall sequence duration accordingly.

Drag this keyframe up or down to adjust the clip speed and to shift the position of the clip's last frame within the sequence.

This keyframe represents the first frame of the clip in the sequence. Dragging it up (making the graph less steep) slows the clip's playback speed. Conversely, dragging the keyframe down (making the graph steeper) speeds up clip playback.

This zoom slider lets you scale the time graph size. Here, the view is zoomed out to show the speed graphs for the clips that precede and follow the one open in the Viewer. Note that you cannot adjust the other clips' speeds unless you open them separately in the Viewer.

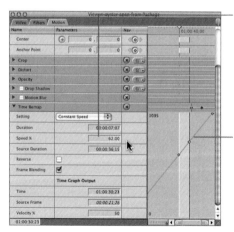

With Constant Speed selected, typing in 62 as the Speed % value slows the existing clip in the Timeline to 62% of normal speed and increases its duration from 4 seconds to 7 seconds. Final Cut Pro applies a ripple edit to extend the duration of the clip (and the entire sequence) in the Timeline.

This graph, with a slope of 45 degrees, in this case prior to accepting the change, shows the 100% speed only because of the granularity of the mini-Timeline. No changes have been applied. Press Return after typing a new Speed % value to apply changes and update the graph automatically.

In this case the entire edited clip has been slowed down to 62% speed. Its duration increased accordingly, as did the overall sequence running time.

The red render status bar indicates that the video and audio need to be rendered for accurate real-time playback. Choose Sequence > Render Only > Needs Render.

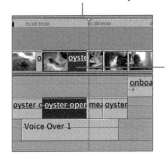

The slowed-down clip takes longer to play, and the ripple edit increases the overall duration of the sequence. To avoid changing the duration of the clip and the sequence, change the clip speed using the keyframes in the Viewer's time graph, rather than the Viewer's Speed % setting.

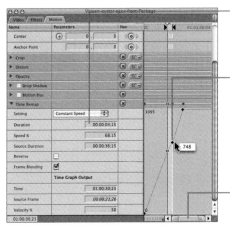

Adjusting playback speed using the time graph keyframes doesn't change the duration of the clip or of the overall sequence.

Drag the second keyframe on the time graph downward to make the clip play back in slow motion; drag it upward to increase playback speed. Here, the keyframe has been dragged down from 100% to 68%, as reflected in the Speed % box.

Adding keyframes for short clips is simplified by zooming into the mini-Timeline with this slider.

126 Basic Fixes

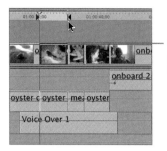

The clip's duration in the Timeline shows that only the speed has changed, much as in a fit to fill edit.

Time Remapping (Variable Speed)

Time remapping has become very popular lately. A clip is playing normally, then it speeds up, then it slows back down. It was very difficult to do something like this in the videotape era, and it's still tricky with many nonlinear editing tools, but with Final Cut Pro, it couldn't be simpler.

After editing the clip you want into the Timeline, double-click it to open it in the Viewer, and click the Motion tab.

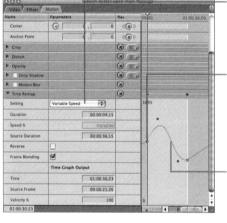

Select Variable Speed from the Setting pop-up. This allows you to create curves, not just straight lines, when you add keyframes to the time graph.

Here you can place keyframes using the Pen tool (press P), or just drag the endpoints to adjust the clip's speed—and the rate at which it changes. Play the clip as you make your adjustments to get a feel for their effects.

The duration in the Timeline/Canvas remains the same. Only the speed at which the clip is played changes.

Fixing White Balance

Sometimes video arrives improperly white balanced. This can occur when a photojournalist forgets to switch filters going from inside to outside, which makes the video look blue. Or if a camera is brought inside from the outside, the video can shift orange. Final Cut Pro makes it easy to fix these problems.

Drag the playhead to any frame of a clip that needs correction, and then choose Effects > Video Filters > Color Correction > Color Corrector 3-way.

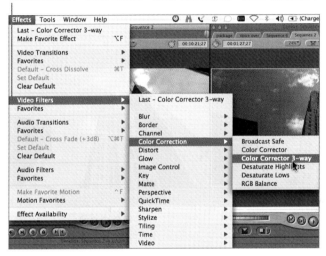

In the Timeline, double-click in the clip that needs correction. This opens that clip in the Viewer.

128 Basic Fixes

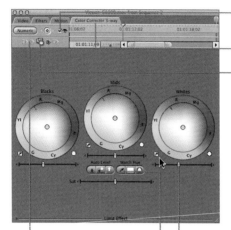

This checkbox turns color correction on and off.

Click the Color Corrector 3-way tab.

Once you fix the color, you can apply the same correction settings to subsequent clips. To do so, select one or more other clips in the Timeline and drag the hand icon to any one of them. You can also save the settings for reuse by dragging the icon to the Favorites bin in the Browser's Filters tab.

Click this eyedropper for the white level and then click it in the clip on a spot that should be white.

When you adjust the white level with the eyedropper, this slider moves to reflect the change.

These numbered-arrow buttons let you apply the color settings found on the clips neighboring this one in the Timeline: Choose among the two clips that precede this one and the two that follow it, or use the sliders to apply custom settings.

 Detail of the eyedropper

TIP In order to "warm up" a shot, try clicking a light blue object in the clip instead of a white one. This can give the image a warm, "Tuscan afternoon sun" look.

Blurring a Subject

There are times when an object in a shot needs to be obscured—a face, a license plate, a billboard, or any other item that might raise legal concerns if it is published and identifiable.

Blurring a Subject **129**

The first thing to do is to isolate the part of the shot you want to blur.

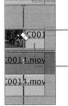 Navigate to the frame just before where you want to add the blur. With the Razor Blade tool (press B), slice the shot into two pieces.

If needed, go to one frame after the subject goes out of frame and split the shot there as well.

 Shift-Option-click the clip and drag it up, without moving it left or right (keep the counter at 00:00:00) and release the mouse button to copy it.

 The duplicate clip in the first video track will act as a background, over which a blurred mask will be applied to selectively obscure part of its contents.

You will drag filters and drop them here, on the second video track.

1 Use two filters to accomplish this effect. First, choose the Mask Shape filter from the Matte folder and drag it to the clip in the second video track.

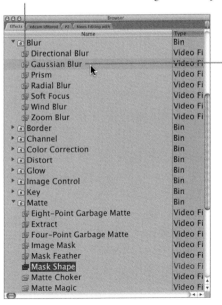

2 Drag the Gaussian Blur filter from the Blur folder to the clip in the second video track. (Note that the blur must be applied *after* the Mask Shape to achieve the correct effect.)

Double-click the clip you dropped the effects on to open it in the Viewer. Click the Filters tab to configure the effects you've just applied.

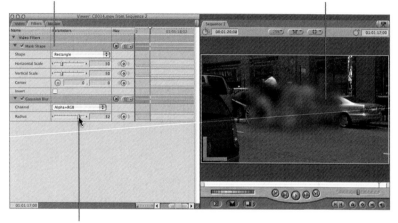

Stacking the filters in this order provides you with a masked blur.

The rectangular mask shape cuts a "hole" that reveals part of the blurred second video track behind the unfiltered first video track.

Using the default crop in the Mask Shape, increase the radius of the blur until you can really see a change in the preview window.

To make it easier the next time, make this filter combination a favorite effect that can be dragged to a clip. All that's required afterward is to adjust the shape and position of the mask.

Blurring a Subject 131

In the Browser's Effects tab, favorites that combine multiple filters are stored in folders. Name the folder what the effect is, in this case a mask blur.

Final Cut Pro stores filters alphabetically and defaults to placing Gaussian Blur above Mask Shape in the folder. If you dropped such a folder on a shot, the effects would be applied in reverse order from what you intended. To correct this, click the Mask Shape filter to highlight its name and add a space before the initial M in its name. This reorders the two filters, placing Mask Shape at the top, so dragging the Mask blur folder will apply the effects in the correct order.

Click the Center parameter point control (+ button) and then locate the center of the shot where you want the blur to begin.

A keyframe is added each time a new center point is assigned to the mask.

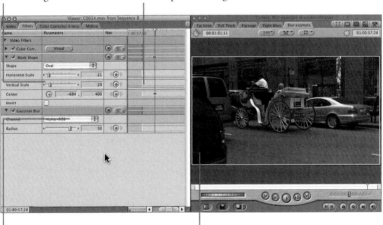

Keyframes can also control the amount of blur. If a subject is moving toward the camera, use the slider as necessary to increase the blur radius to keep the subject unrecognizable.

The shot pans from left to right, and blurring this license plate requires keyframing the movement of the mask. The red plus sign indicates the center of the mask.

Scrub through the clip, watching for when the image peeks out from under the mask. When it does, recenter the blur by dragging the crosshairs, which also adds a keyframe.

Continue scrubbing through the shot and make sure what needs to be blurred is never visible from under the blur.

Continue through the end of the clip, recentering the blur as you go. Keyframes are added.

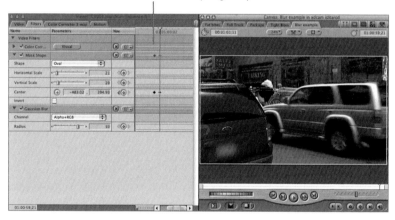

Motion 3, a component of Final Cut Studio 2, also includes a variety of blurs that can be applied to a shot in the Final Cut Pro Timeline. Please consult *Apple Pro Training Series: Motion 3* for details on this.

Highlighting a Subject

It's sometimes desirable to highlight someone or something in a clip to direct viewers' attention. It could be a person in a crowd, a football being fumbled, or words in a legal document. The technique for doing so is effectively the inverse of the one used to blur an object, as it essentially entails dimming everything in the image except the part you want to highlight. So a mask is once again involved, but its *alpha channel* is reversed to apply it to everything in the foreground frame except the area designated by the mask.

Shift-Option-click the clip you want to highlight and drag it upward to the first video track, as you did in creating the blur.

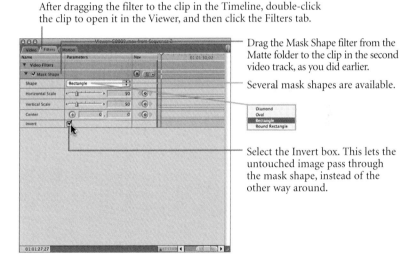

After dragging the filter to the clip in the Timeline, double-click the clip to open it in the Viewer, and then click the Filters tab.

Drag the Mask Shape filter from the Matte folder to the clip in the second video track, as you did earlier.

Several mask shapes are available.

Select the Invert box. This lets the untouched image pass through the mask shape, instead of the other way around.

You will need to add the Brightness and Contrast filter to the effect in order to lighten or darken the surrounding area. There will be two filters in the Viewer detail: Mask Shape, and Brightness and Contrast.

As with animating the blur, use the Center point control (+) to recenter the highlighted area. Keyframes will be added as you move the highlight.

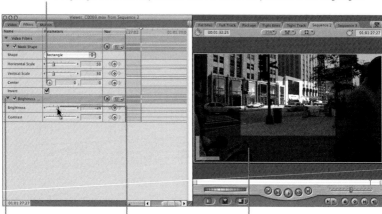

Drag the Brightness filter from the Image Control folder to the Viewer or to the same clip in the Timeline.

Lower the brightness to dim the unmasked portion of the image.

To highlight this car, center the mask on the car, reduce its size, and then reposition the mask's center as you scrub through the clip to keep the highlighted window on the car as it moves.

Resize

Sometimes it's useful to enlarge a clip slightly to push its edges out of frame to exclude a station or network logo, a microphone boom, or to otherwise improve the composition of a shot. This is easy to do in Final Cut Pro. Gradually enlarging a shot over time is also a good trick for simulating a camera zoom, but do so with care because image quality decreases as the enlargement level increases.

Resize   135

Double-click the clip in the Timeline to open it in the Viewer, and then click the Motion tab.

This image is at full 100% screen size.

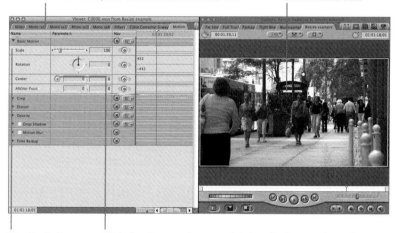

Use the Scale slider to adjust the size of the image.

Click the Center point control (+) to display crosshairs that specify a point on which a zoom will recenter the frame. Keyframes are assigned each time a new center point is selected. This allows a "limited" zoom, mimicking a camera move, which can add interest to otherwise static scenes.

In this case, a 74% increase in the picture size results in the composition on the right.

By adding keyframes here with the Pen tool (P), you can animate a zoom.

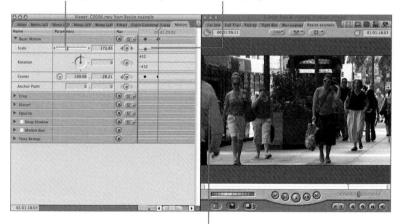

Note the difference in composition after the 74% zoom from the previous image.

Adding Other Effects

A full exploration of Final Cut Pro's many effects is beyond the scope of this Quick Reference Guide. For more details on the variety of effects and how to achieve them, please refer to any of the Apple Pro Training Series books on Final Cut Pro 6. A brief example of one effect, Spinback 3D, is provided to get you started.

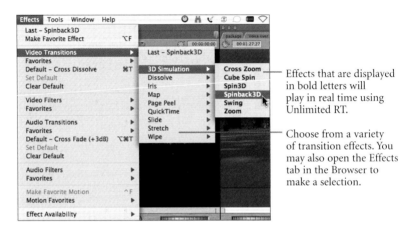

Effects that are displayed in bold letters will play in real time using Unlimited RT.

Choose from a variety of transition effects. You may also open the Effects tab in the Browser to make a selection.

The same effect is now not bold as a result of switching from Unlimited RT to Safe RT in the Timeline. This effect will require a render while the Timeline is set to Safe RT. There are some simpler effects that could be played in real-time when in Safe RT, but it is dependent on the type of effect, available processor power, and the format and compression of the clips being affected.

To open up the parameter details in the Viewer, double-click the transition icon in the Timeline.

All effects open with a display like this in which you can change the various parameters of the effect. Here you can adjust the position of the effect (at the start, center, or end of the transition).

In this case the adjustable parameters are the border size, color, and the angle of the axis of the spinback.

Image Stabilization—SmoothCam

SmoothCam is a plug-in that has migrated to Final Cut Pro 6 from Shake. It analyzes a clip and lets you apply varying levels of correction once the analysis has been made. More detail is available in *Apple Pro Training Series: Final Cut Pro 6*, by Diana Weynand. A brief example is described in this section.

Highlight the clip in the Timeline you want to analyze and choose Effects > Video Filters > Video > SmoothCam.

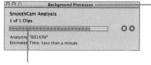

Analysis is a background process. You can select multiple clips for analysis. They will be analyzed one at a time in the order you select them.

A progress bar tracks the SmoothCam analysis. Also, a yellow clip overlay that reads "Analyzing Clip" appears in the Canvas while the analysis is being performed on the clip.

Once the analysis for a clip is complete, double-click it in the Timeline to open it in the Viewer. Click the Filters tab to see the SmoothCam filter settings. The correction has already been applied, so view the adjusted clip prior to trying out any small adjustments.

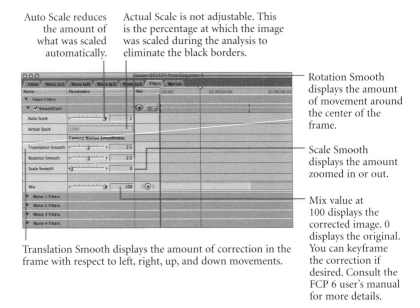

Auto Scale reduces the amount of what was scaled automatically.

Actual Scale is not adjustable. This is the percentage at which the image was scaled during the analysis to eliminate the black borders.

Rotation Smooth displays the amount of movement around the center of the frame.

Scale Smooth displays the amount zoomed in or out.

Translation Smooth displays the amount of correction in the frame with respect to left, right, up, and down movements.

Mix value at 100 displays the corrected image. 0 displays the original. You can keyframe the correction if desired. Consult the FCP 6 user's manual for more details.

Audio Equalization

On occasion you will need to clean up some audio to make the sound more clear and understandable. This can be done with the built-in audio EQ filters, as well as with Soundtrack Pro 2, the companion application to Final Cut Pro 6 found in Final Cut Studio 2. The built-in audio EQ can handle some of the requirements, but Soundtrack Pro has extensive capabilities. We will address only Final Cut Pro in this section.

Identify a sound bite that needs EQ to make it more understandable.

An example could be a sound bite with an air conditioning compressor overwhelming an interviewer's question, or fire engine pumps drowning out other sound needed in the story.

Audio Equalization 139

Choose Effects >
Audio Filters >
Final Cut Pro >
3 Band Equalizer.

Double-click the audio track you want to equalize to open the clip in the Viewer, and then click the Filters tab.

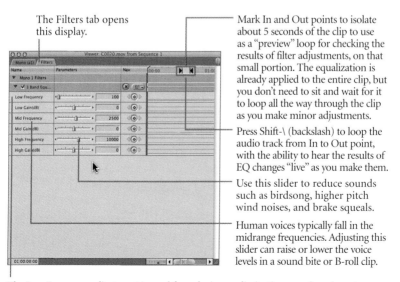

The Filters tab opens this display.

Mark In and Out points to isolate about 5 seconds of the clip to use as a "preview" loop for checking the results of filter adjustments, on that small portion. The equalization is already applied to the entire clip, but you don't need to sit and wait for it to loop all the way through the clip as you make minor adjustments.

Press Shift-\ (backslash) to loop the audio track from In to Out point, with the ability to hear the results of EQ changes "live" as you make them.

Use this slider to reduce sounds such as birdsong, higher pitch wind noises, and brake squeals.

Human voices typically fall in the midrange frequencies. Adjusting this slider can raise or lower the voice levels in a sound bite or B-roll clip.

The Low Frequency adjustment is good for reducing or eliminating sounds such as air conditioning compressors and low wind noises that might reduce the clarity of a sound bite.

If you have more than one clip that requires the same audio EQ settings (such as, for instance, excerpts from the same source clip), it is easy to duplicate adjustments and assign them to other clips in the Timeline.

1 Highlight the clip with the corrected EQ and press Command-C to copy.

2 Command-click the other clips you want to adjust.

3 Press Option-V to open the Paste Attributes window.

4 Select the Filters box and then click OK.

Soundtrack Pro, a component of Final Cut Studio 2, provides many additional audio features, many of which are specifically designed for video production.

There is a wealth of effects built in to Final Cut Pro, as well as many plug-ins developed by third parties. For more information on what is available and how to apply and adjust many of these effects, please consult Peachpit Press's Apple Pro Training Series books *Final Cut Pro 6* and *Advanced Editing Techniques in Final Cut Pro 6*.

9
Delivering the Story

The final stage of editing a project in Final Cut Pro is delivering a finished story to your audience.

If that audience will see your story on TV, your choices are transferring the project file to a video server; playing the contents of the project Timeline as video and audio from your Mac to a video server for playback; or playing directly to air from the Final Cut Pro Timeline.

At a newspaper or Web news service, and at broadcast outlets that repost stories online, your story must be *transcoded*, or converted into a Web deliverable.

This chapter covers each of these delivery options.

Preparing for Air

When preparing a project for broadcast, your choice of delivery method largely depends on the playback solution implemented at your station or network.

Detailed descriptions of server-based production and playout are beyond the scope of this book. There are too many variables in server-based systems, and too many customizable options to cover in one slim volume. For details about your specific server-based setup, seek on-site guidance from your engineering department.

The instructions below address broadcast environments in which video is delivered via playback from the edit room, or via field playback from a MacBook Pro.

The Final Cut Pro Timeline can play back edited compositions live to air using a number of methods. These include output to third-party graphics cards such as those from AJA and Blackmagic Design; output to an external DVI converter box such as the Matrox MXO; and output to a FireWire converter box, such as the Canopus ADVC series and those from AJA. You can also play out via most FireWire-enabled broadcast and consumer cameras and decks.

Safe RT, Unlimited RT, and Rendering

Final Cut Pro's ability to generate effects and transitions as you play back a sequence depends on two things: the system hardware and the number and complexity of the effects and transitions. Processor type and speed, amount of installed RAM, and graphics subsystem type all interact to determine whether a particular Mac can generate a sequence's effects and transitions on the fly, in real time, as you play it back.

Among the items to consider for real-time playback of effects are the following:

Hardware

- ▶ Processor: The faster and more powerful the processor, the better the real-time effects performance.
- ▶ Amount of RAM: More RAM means better performance.
- ▶ Storage-access bandwidth: Faster access to local drives or networked storage is better.

Software

- Number of applications open and running. (Fewer is better.)
- Number of open projects and open sequences in the Timeline. (Fewer is better.)

Sequence content

- Number of effects and transitions. More effects place greater demand on the Mac.
- Number of tracks in the active sequence. Overlapping transitions and effects make heavy demands on system hardware.
- Number of audio tracks in the active sequence. More audio creates more work for the Mac.
- Compression of source material. The long-GoP standard, native to HDV and XDCAM HD material, is especially processing intensive.

If a transition or effect is too complex for the Mac to generate accurately, Final Cut Pro can approximate it for preview purposes, cutting corners on image quality while still giving you a good idea of how your sequence will look. For final output, however, those transitions must be *rendered* in order to play back in real time at full quality. Rendering can often be a slower-than-real-time process that builds new frames, with effects and transitions applied, and transparently inserts them into your sequence.

To guarantee a successful playback, you should render every effect and transition in your sequence, but when you're working on deadline, there are a few shortcuts you can take to avoid rendering everything. Knowing which effects must be rendered and which can be generated on the fly requires use of the Timeline's Safe RT and Unlimited RT modes.

The RT (real time) modes provide an easy-to-use gauge of real-time performance, specific to the Mac that's being used to play back the sequence. To activate the RT modes, click the RT pop-up menu button in the upper-left corner of the Timeline window.

Click to open the Real Time
Effects (RT) pop-up menu.

When Safe RT is selected, Final Cut Pro plays only those effects and transitions that it can without dropping frames or reducing picture quality.

When Unlimited RT is selected, FCP does a "best effort" playback of all effects and transitions, reducing picture quality and frame rate as necessary to keep the playback time-accurate. This effects "preview" can be useful in testing the clip's placement and pacing without having to render it to see it.

Menu options in the Playback Video Quality and Playback Frame Rate sections let you control how Final Cut Pro cuts corners when approximating effects in Unlimited RT mode. Select Dynamic for both settings to let FCP make its best attempt at all effects.

When working on deadline, you should always leave your Timeline in Safe RT mode so you can quickly see what needs to be rendered. When all effects are rendered, the Timeline should play out with no problems.

Along the top of the Timeline, just above the ruler, are two *render bars*. The top is the video render bar; the bottom is the audio render bar. As you add audio and video effects or transitions to your Timeline, you will see different colors appear in the video or audio (or both) render bars to indicate whether a given effect can play back in real time on your Mac.

A red render bar means this clip must be rendered when the Timeline is set to Safe RT. The "safe" means you won't drop frames because of the processor. Light green indicates a real-time preview effect. It will play in real time without dropping frames but it won't play at highest quality. A dark green status bar indicates that an effect or transition will play back in real time without any problem.

Select Safe RT. Safe RT ensures that no effect will be played that exceeds your processor's capacity to play without dropping frames.

As you add more video or audio tracks, the playback complexity increases.

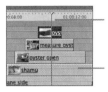
This sequence is in Unlimited RT mode, with frame rate and video quality both set to Dynamic. The yellow render bar means the sequence will play, but with compromised quality.

5 video tracks, all created using the picture-in-picture effect

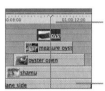
A switch to Safe RT, which does not allow playback of effects or transitions that can't be generated in real time, turns the render bar red. The sequence must be rendered in order to play back accurately in real time.

Same sequence as in previous illustration

When you choose Sequence > Render All > Both, the selected audio and video track regions in this menu will be rendered. Toggle the selection for a color entry by choosing it in the menu. Leaving the options selected as they appear here works well for most news and sports applications.

You don't have to worry about rendering effects or transitions that display dark green render bars in Safe RT mode. For best results before playback to air, you should render all effects and transitions that display any of the other render-bar colors shown in this menu.

A blue render bar denotes a sequence that has been rendered and is ready for full-quality playback.

In many cases, dissolves won't need to be rendered for playout from the Timeline.

The orange bar means the sequence will play in real time when set to Unlimited RT. For playback to air, the sequence should be rendered after the setting is switched to Safe RT.

Unlimited RT will ramp the picture quality and /or frame rate up or down when set to Dynamic.

While editing news under a deadline, you should always have Safe RT on. This will prepare you for what will ultimately require rendering as you get closer to air. You may choose to render as you edit in order to give yourself more creative time as you approach deadline.

If you are editing a particularly effects-laden sequence and want to check it for shot placement and pacing, you should set the Timeline to Unlimited RT mode and set Playback Frame Rate and Playback Picture Quality to Dynamic. This will provide you with immediate creative feedback while editing.

As soon as you've got your clips ordered and adjusted for duration, switch back to Safe RT mode. When finished, choose Render All to render the Preview RT effects and transitions.

If you are editing on an Xsan or other server, you should render your effects directly to shared storage, not to your local hard disk. Choose Final Cut Pro > System Settings or press Shift-Q to open the System Settings window. Click the Scratch Disks tab and then click the topmost Set button. In the Browser that appears, navigate to your shared storage volume. Click Choose in the Browser and then click OK in the System Settings window.

Live to Air

When it's time to deliver a sequence for broadcast, one option is to play it from the Timeline direct to air. If your edit room is wired to the control room and your Mac is equipped to output broadcast-ready video and audio, the process is simple.

You will need some form of hardware to get video and audio off the Timeline. You could use external FireWire converters, such as those from Canopus and AJA, or a comparable internal video card in the Mac Pro. Most professional broadcast cameras and decks enabled with FireWire, as well as prosumer camcorders, can also be used to convert the signal off the Timeline into video for transmission.

Patch the video outputs from the converter or video card into inputs at the station—a procedure roughly comparable to connecting a DVD player to a TV—and then do the following:

1   Confirm your Safe RT settings.

2   Confirm that the required effects will be rendered, by choosing Render All.

3   Cue the sequence in the Timeline.

4   Wait for the instruction to roll from the control room.

5   Click the Play button or press the spacebar.

This method works for playing video to air both in server-based editing environments and on standalone editing workstations storing video on their internal drives.

Playback to Tape

If you will be playing the edited sequence back to videotape, you will need to make sure your connections to the deck are correct, and then choose Edit to Tape from the File menu. If you use this command often, it's a good idea to assign it to a button above the Timeline or to create a keyboard shortcut for it.

When you do so, you're given the option to generate color bars and tone, add a slate that is built in, or play a file you may have created. You can also add the story slug to the leader content and add black at the end of the sequence. You have the option of putting it onto a pre-timecoded tape at a specific point, or blasting it into assemble mode, creating a new timecode reference on the tape.

As always, the number of effects that play in real time will determine how much will need to be rendered prior to output. In Safe RT mode, you may need to render portions of the Timeline.

> **NOTE** ▶ When playing an XDCAM HD or HDV Timeline to tape in Edit to Tape mode, the system will automatically force the creation of a new file as a result of the frame structure of HDV. A workaround for this is to "crash record" the tape and play it in the Timeline. What you lose with this method is the ability to drop the story directly to a specific timecode value on the tape.

Remote Story Delivery

Depending on your location and situation, the logistics of getting video or a file back to the home office varies. From accessing satellites orbiting Earth 22,000 miles away to using cellular phone towers, each method has multiple techniques that result in a wide range of quality.

Playing the Timeline to a Sat Truck

Current microwave and satellite trucks employ encoders that accept video signals and transmit them over a point-to-point radio link. These encoders are produced by various manufacturers, and each has its own setup requirements.

Encoders accept standard definition video via either composite NTSC/PAL or Serial Digital Interface signals. For high definition video they accept either HD-SDI or Asynchronous Serial Interface signals, or both. The SDI and ASI signals include audio tracks embedded with the video. NTSC requires an additional analog audio connection.

When editing SD video such as DV25 or DVCPRO (at 25 Mbps), a FireWire converter such as those made by Canopus and AJA enables you to play the contents of your Timeline to the encoder in the truck. In most cases you will have to render all effects and transitions for output via a FireWire converter except those that are dark green in the Timeline.

With HD video, playing back an edited story as video is a little more complicated.

If you're working with DVCPRO HD video, you can use your deck or camera to make the conversion.

1 Run a six-pin FireWire cable from the DVCPRO HD deck to the MacBook Pro, and use the deck's hardware controls to set its input to *1394*.

2 Use a BNC video cable to connect the deck's HD-SDI output to the truck's microwave encoder.

3 Activate the deck's E-to-E function and then play the Timeline through the deck to the microwave encoder.

If you're using a Sony XDCAM HD, this trick won't work. The XDCAM's long-GoP MPEG compression format prevents live playback over a FireWire port unless you completely render the entire sequence and all the effects into a new file.

The alternative is to use a Matrox MXO converter box, which allows you to play out XDCAM-native HD video (as well as other HD formats like DVCPRO HD), including simple effects, dissolves, and color correction, in real time.

If you need to, you can also use Matrox MXO to downconvert and output XDCAM-native HD sequences to a standard definition microwave encoder.

Set up the Matrox MXO as follows:

1 Connect the special Matrox MXO cable to your Mac.

The end of the cable with four connectors plugs in to the Mac. Insert the DVI connector into the Mac's DVI port, its USB connector into one of the Mac's USB ports, and its male stereo mini-plug into the Mac's audio-out/headphone jack. You can connect headphones or external speakers to the fourth connector, the female stereo mini-jack, but that's not necessary for playback.

2 Plug the two-connector end of the special cable in to the Matrox MXO. Insert the wide connector into the DVI "in" connector, and the USB "B" plug to its connector.

3 Run a BNC video cable from the Matrox MXO's HD-SDI port to the microwave encoder.

4 Plug in the Matrox MXO power supply. A blue light should appear, confirming the power source.

5 If you haven't loaded the Matrox software yet, do so now.

If this is the first time you are installing the Matrox software, you will need to restart the computer.

Once this is done, you will see the computer screen fade and come back as the system senses the second screen. Open System Preferences, choose Displays and turn mirroring off if it is on.

Click the new Matrox MXO icon in the main System Preferences, and then click the Gather Windows button on its Display pane. A second preferences pane, titled Matrox MXO, will appear on the screen.

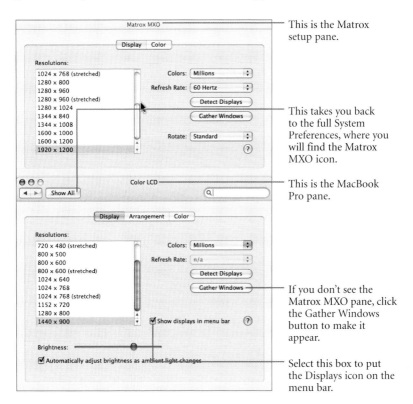

This is the Matrox setup pane.

This takes you back to the full System Preferences, where you will find the Matrox MXO icon.

This is the MacBook Pro pane.

If you don't see the Matrox MXO pane, click the Gather Windows button to make it appear.

Select this box to put the Displays icon on the menu bar.

Click Show All to display all the System Preferences. At the bottom is Matrox MXO. Click to open the preferences and to set up the MXO to your requirements.

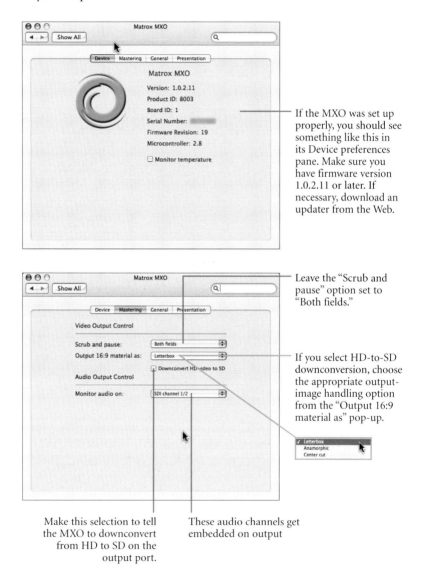

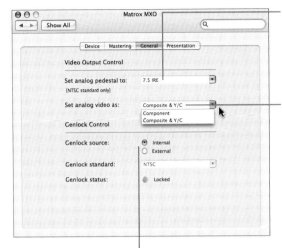

Leave at 7.5 IRE, unless in Japan or Europe, where the correct setting is 0.

The "Set analog video as" pop-up lets you choose between Component or Composite & Y/C (Y output). Most sat and microwave trucks accept either Composite or SDI. Leave on Composite unless you are making a dub to a Betacam deck with the component outputs.

The "Genlock source" option defaults to Internal, but you can feed it a reference signal and choose External. Choose the correct signal type from the "Genlock standard" pop-up.

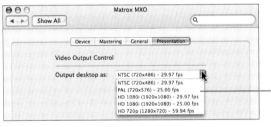

These are the currently supported outputs from the MXO.

You may also need to adjust the video output resolution in the Presentation pane of the Matrox MXO preferences, but the default setting of 1920 x 1080 pixels works well for most HD material.

If you are, for instance, editing with Final Cut Pro with native P2 SD or HD material, or XDCAM SD or HD material at 18, 25, or 35 Mbps, you may also add a small number of real-time transitions and simple effects and play the Timeline live to air using the Matrox MXO. This magically eliminates the "recompile of the GoP" for XDCAM HD material that most other editing systems require.

To do this, finish editing your sequence and check the real-time performance in Safe RT, rendering any suggested portions of the Timeline. A dark green bar denotes the only real-time effect display that ensures highest-quality playback.

Make sure the Timeline is set to the same format and frame rate as the original material. Go to the View menu and choose the MXO HDV 1080i 29.97 YUV from the Video Playback submenu, as shown below.

For P2 and Panasonic DVCPRO HD tape that was ingested and edited, you can choose from several DVCPRO HD options. Note the pixel dimensions for the 720p settings: 960 x 720. DVCPRO HD at 1080i has pixel dimensions of 1280 x1080.

To play back XDCAM HD through the MXO, choose the HDV 1080i 29.97 (or 25 for PAL) and YUV. You need to make sure that you have selected 1440 x 1080, as that is the pixel dimensions of XDCAM HD.

Then, with your lone BNC connector going to your HD microwave encoder carrying embedded audio and uncompressed video, choose View > External Video > All Frames or press Command-F12.

The MXO will now take over the DVI output and treat it like a second screen in an edit room. It then delivers the uncompressed (temporally recompiled GoP) video directly off the Timeline, with audio to your microwave encoder.

If you need to send SD to the encoder, click the Matrox icon at the bottom of System Preferences. Click the Mastering tab and choose "Downconvert HD video to SD," and close the window.

Broadband Story Delivery

With the wide availability of wired and wireless broadband Internet, it is possible to send your story back as a digital file, without playing video. The process of creating this file is called *transcoding*.

This involves compressing your edited sequence into a relatively small digital file, balancing the possible loss of image quality against the time it will take to make the file and transmit it back via several delivery methods.

Here are some numbers to refer to as you read the following section:

- 1 megabyte (MB) equals 8 megabits (Mb)
- 1000 kilobits (Kb) equal 1 megabit (Mb)
- One minute of DVCAM and DVCPRO 25 (captured at 25 Mbps or 3.125 MBps) yields a data file of just under 200 MB.
- One minute of DVCPRO HD (100 Mbps) consumes about 800 MB of storage capacity.

- A typical home DSL line allows data transfers of 50 Kbps, or roughly 3 MB per minute. At that rate, it would take more than an hour to send one minute of uncompressed DVCAM or DVCPRO 25 video.

As an example, the package edited in Chapter 7 runs 1:22, a typical news package length, and is 300 MB (2400 Mb) in size.

Compressed using the H.264 codec, set to high quality for iPod delivery (640 x 480 pixels, which comes fairly close to the 720 x 486 pixel dimensions of an NTSC broadcast signal), the file size shrinks to 16 MB. Or you can reduce the picture quality to the regular iPod setting at 320 x 240, or one-quarter the screen size, and use up only about 6.5 MB.

Compressing the sequence and sending it back via broadband is great for remote locations where you can't get to a feed point or truck.

It is a good idea to work out the logistics of this delivery method before having to use it under pressure. There is nothing worse than having crashed a story for air and not being able to get it back in time.

The station or bureau needs to set up a secure FTP (File Transfer Protocol) server. This provides a secure path for copying digital files over the Internet, with error checking to prevent the transfer of incomplete or corrupted files. It's critical to have the necessary login and password information on hand (or saved in your FTP software) when you begin an FTP transfer.

In most cases, you'll use the built-in FTP function of Compressor, included with Final Cut Studio, to transfer your files. But if your station doesn't take that route, use an FTP utility such as Softworks' Fetch or Panic's Transmit, which allow file transfers to pick up where they left off in case your Internet connection is lost. This is critical when you have 85 percent of the story fed and your signal drops. You then resume the transmission, and the story is stitched together seamlessly on arrival.

There are several types of connections by which you can send your edited story.

- Typical residential or business DSL (from the phone company)
- Typical cable modem
- Broadband EVDO or UMTS card from cellular companies
- Cell phone with 3G service (using EVDO or UMTS) and a tether or Bluetooth connection
- BGAN (Broadband Global Area Network) satellite uplink
- Wireless access points such as those found at airports and cafés

These connections have their strengths and drawbacks. For instance, DSL from the phone company is referred to as a "home run" to the Central Office (CO). This means that a pair of copper wires connects at your home or office, and with no interruption or further connection, terminates at the CO. This is a strength. The cable companies have various methods of distributing their Internet connections. Typically there is a box located in a neighborhood that aggregates the connected customers and sends the combined signal down a fiber-optic circuit back to the cable headend.

DSL

A DSL (digital subscriber line) connection provides stable upload speeds, rated at about 400 Kbps. Stable is the key word. You can count on this speed.

Cable Modem

Cable modems are analogous to what were referred to in the 1940s and '50s as "party lines." A cable-modem connection is typically shared by many neighbors. If you are the only one in the neighborhood with a cable modem connection to the Internet, you will get tremendous speeds. As each of your neighbors connects, depending on your cable company's architecture, your available bandwidth could be cut in half with each new connection. You are sharing the path back to the cable company for your neighborhood with any and all that connect on that node. For a 5 p.m. TV station deadline, cable modems might not be good because school is out and many users could be using the available bandwidth for studies (and gaming). Your performance will fluctuate at different times of the day and night with cable modems. There is no "guaranteed" performance with cable modems.

EVDO and UMTS Wireless Cards

EVDO (Evolution Data Optimized) and UMTS (Universal Mobile Telecommunications System) wireless broadband rely on cellular-phone technology and deliver top transfer speeds that rival DSL's, but they also have a shared architecture. Multiple users share a connection to a cell phone tower, so the more activity on a connection, the less bandwidth is available. This is a major consideration at news events where many crews are fighting for bandwidth. Also, cell tower backup batteries typically fail when electrical power is out for more than 36 hours, making EVDO and UMTS transfers impossible.

An EVDO or UMTS card can be an external USB connection to a laptop or desktop, or a card that slips into a laptop. You connect with software much like the old days with dial-up telephone modems.

3G Service with Cell Phones

For the same reasons described above for EVDO and UMTS connectivity, the use of a cell phone for file transfers at a major news event is also an option. Many 3G (third generation) cell phones rival EVDO/UMTS speeds, and when connected to your Mac via Bluetooth or tethered with a USB cable, can be used for data transfers. A USB tethering cable will provide better results than a wireless Bluetooth connection.

Broadband Global Area Network

Satellites administered by Inmarsat, an international consortium, provide global communications links for ships at sea, users in remote areas, and news organizations.

Several manufacturers make BGAN uplink hardware. These can be deployed in the field when no other communications method is available. BGAN transmitters are battery operated, which makes them suitable for areas where power is out after a disaster. A BGAN connection provides both voice and data capabilities. Each BGAN has its own phone number and is used in a similar way to early telephone dial-up methods, but is faster.

BGAN transfer rates range from 56 to 256 Kbps (versus consumer DSL upload speeds of 400 Kbps), so it's a good idea to get familiar with an individual transmitter's speed capabilities before you try to use it on deadline.

Note also that obtaining a BGAN connection can be tricky during coverage of major news events, where many organizations are likely to be competing for a limited number of satellite "circuits."

Wi-Fi Connections

Wireless connections, as a result of their popularity, will most likely provide you with the lowest transfer speeds of any described here. An airport or café wireless connection is shared with the rest of the people at that location.

Wi-Fi technology, or 802.11 wireless networking, comes in several "flavors," each with a different theoretical maximum connection speed. The most common flavor, 802.11b, is rated at 11 Mbps; the 802.11n flavor, supported in the newest MacBook Pro, provides a theoretical transfer rate of 108 Mbps, but only at access points that support it.

This is only the connection speed you have to the antenna, not the connection out to the Internet.

Many commercial hot spots still offer only 802.11b access, which forces 802.11n devices to dial back to 802.11b speed. In addition, if ten people are sharing a public access point, your transfer speed can be reduced by 90 percent. There is no guaranteed performance with Wi-Fi.

Compressing the Story

Each of these methods requires that your story has all video and audio effects rendered, and that you export your story in a compressed format to maximize the data throughput during the time leading up to your deadline.

Once you finish rendering the story, you need to compress it for transmission.

For best results when transferring digital files, compression options, which include QuickTime, MPEG-1, MPEG-2, MPEG-4, and H.264 (used for iPod video) should be tested and approved by the station, and then configured in the Compressor encoder before use.

As an alternative, the engineering department can set up a Compressor *droplet*—an icon representing one or more compression presets that can be placed on the desktop for easy access. Just drag an FCP project onto a droplet to compress it, without even opening Compressor. If unsure of the correct settings to use, ask the station to email a droplet.

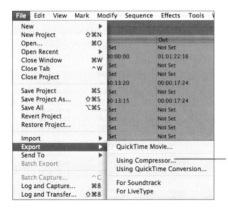

In Final Cut Pro, choose File > Export > Using Compressor to open Compressor and begin compressing the sequence in the Timeline.

Final Cut Pro 6 will open Compressor 3 and load the selected sequence in the Batch window.

The story comes over from Final Cut Pro and loads in Compressor's Batch window, with the default compression and destination settings applied.

Click here to open the Batch Monitor to view the progress of the encoding.

Scrub through the sequence to see how it will look after the selected compression setting is applied.

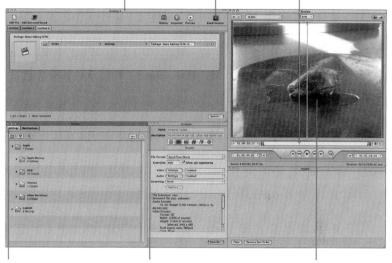

The Settings pane displays preset compression configurations and lets you create custom settings.

The Inspector pane lets you modify settings for video and audio, including frame size and frame rate.

Compressor's Preview window displays "before" and "after" images (on either side of the draggable divider) that show the effects of the selected compression setting before you apply it.

If you want to set a default for the compression setting, open the Compressor Preferences window and select Default Settings.

To make iPod 320 x 240 your default compression setting, click the Default Setting pop-up and choose Apple > Apple Devices > H.264 for iPod Video 320x240 (QVGA).

If you use Compressor for FTP, its values should be set for you, but in case you need to add compression settings and destinations, here is a brief overview of its Settings window.

If you want to save multiple similar compression settings for reuse, click the Settings tab to create a new settings group.

When you highlight a setting or settings and click this button, Compressor creates a "droplet" you can place on your desktop. Drag a project icon to the droplet to apply one or more compression settings at once.

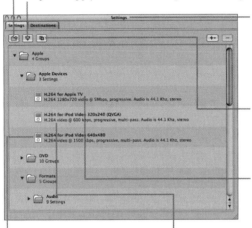

The Settings window displays all the presets already loaded in Compressor. Scroll down to see the Custom folder. This stores any new presets you create, including modifications to existing settings.

Click this button to duplicate settings. This enables you to make easy modifications to customized compression tasks based on a preset.

Use this setting, designed to format files for Apple TV HD, for HD-formatted digital files.

This setting, which formats video at 640 x 480 pixels for use on an iPod, offers image quality close to that of a standard NTSC or PAL broadcast signal.

Files compressed using the iPod at 320 x 240 setting will be about 40% the size of the higher-quality 640 x 480 iPod files.

Remote Story Delivery 163

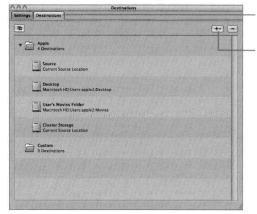

The Destinations tab shows all of the preset destinations.

This button lets you add destinations and opens a new destinations dialog.

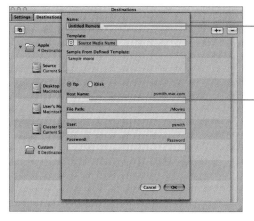

Name the setting something you can remember easily. This could be your default setting for FTP delivery.

This information is typically set up already. Each heading needs a value. The destination "folder" is the file path.

164 Delivering the Story

You can configure the delivery of the files to that particular FTP site so that you don't need to change things each time you are connected.

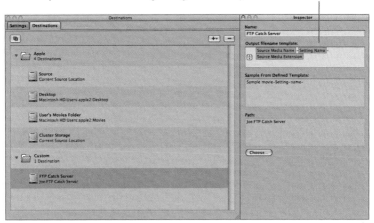

The Inspector window lets you monitor the settings as well as modify them for the specific requirements of an encoding.

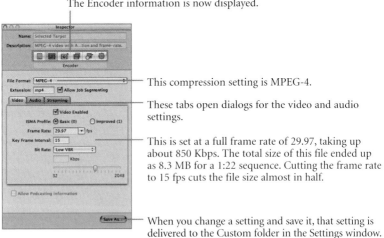

These buttons provide different views in the pane below. The Encoder information is now displayed.

This compression setting is MPEG-4.

These tabs open dialogs for the video and audio settings.

This is set at a full frame rate of 29.97, taking up about 850 Kbps. The total size of this file ended up as 8.3 MB for a 1:22 sequence. Cutting the frame rate to 15 fps cuts the file size almost in half.

When you change a setting and save it, that setting is delivered to the Custom folder in the Settings window.

Once you click Submit in the Batch window, a dialog opens for you to confirm the name of the new file.

Rename the file here. High, Medium, Low priorities are available.

If you are on a network with other computers set up as a cluster, you can use some of their available processing power to accelerate the encoding.

When you click Submit, you can view the progress by opening the Batch Monitor. To do so, click the button at the top right of the main Compressor window.

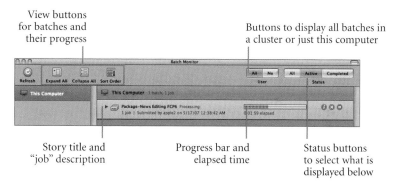

View buttons for batches and their progress

Buttons to display all batches in a cluster or just this computer

Story title and "job" description

Progress bar and elapsed time

Status buttons to select what is displayed below

In some cases you may be delivering directly to an FTP site from Compressor. In other cases you will have the compressed file delivered to the desktop.

Once the file is complete on the desktop, make the connection to your FTP server and drag the file to the Put icon or to the appropriate window for delivering the file.

With Compressor you may add several types of compression to the list. You may also add different stories to the list, and they will get treated sequentially.

As you do more of these deliveries, you will determine not only the best combination for the compression/time ratio, but also the quality of the available broadband connection. Going to a coffee shop when lots of people are using their laptops will not provide you with a level of service that will get your story back quickly. Chances are a quiet coffee shop will have more bandwidth available on its wireless system than a busy one.

Sample Compression and Transfer Speeds

If you're planning to submit a story as a digital file on deadline, you'll need to factor in the time required to compress the file and then transfer it over whatever type of Internet connection you'll be using. Choosing a high degree of compression will reduce the transfer time (and the image quality), but will increase the time required for the compression phase.

For an idea of some of the potential tradeoffs and time considerations, consider the package we created in Chapter 7, "Fast Package Editing." The story running time is 1:22, and its file size, in native DVCPRO 25 format, is about 300 MB.

The table on the next page shows compression and transfer times required to apply two Compressor full-frame-rate presets optimized for iPod delivery, both of which in many cases are sufficient for broadcast use, and which provide highly compressed files well suited for broadband delivery. These settings also generate files that can be delivered straight to a webmaster for inclusion on the website. These results were achieved on a 2.16 GHz Intel Core Duo MacBook Pro with 2 GB of RAM. Results will vary with hardware configuration, sequences of different types, and transfer-connection circumstances.

Compressor setting	Compressed file size	Time (min:sec) required to compress from DVCPRO 25	FTP transfer time (min:sec) via DSL	FTP transfer time (min:sec) via EVDO
iPod 320x240 @29.97 fps	6.5 MB	5:40	2:34	3:30
iPod 640x480 @29.97 fps	15.9 MB	11:45	7: 40	10:05

Again, compression results may vary depending on the source material, processor speed, available RAM, and the number of applications running concurrently. The transmission speeds are all dependent on the type of connection and the competition on that circuit for bandwidth. There are no guarantees in speed of delivery.

Delivering to a Website

Determine the format required for your website. In some cases, both HD and SD material are being made available. Once you have a format for each, create settings and add them to the Compressor Custom folder. If there is one particular setting that is always used, make that the default so that when you export from Final Cut Pro, that setting is automatically set.

The same is true for the destination. If you are working in-house, with no FTP required, your Final Cut Pro system can mount the specified target volume. This can be made your default destination.

If these two settings won't ever change, lock them in as defaults. This presets the compression and delivery, automating both when you click the Submit button.

In some cases you may need to add a podcast or another compression type, and possibly one or two more delivery locations.

Delivering a Video Podcast

You can easily add a setting for both regular and higher-quality iPod deliveries. This is done by opening the Settings tab and opening the Apple Devices folder.

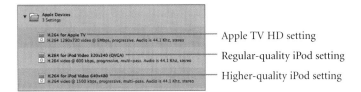

— Apple TV HD setting
— Regular-quality iPod setting
— Higher-quality iPod setting

Grab the setting you want and drag it to the blue area of the Batch window. This adds that setting to the same story as a second "job." If you want to replace the existing setting, drag the new one on top of the one that is already there.

Multiple deliverable encode settings from the same sequence have been dragged to the Batch window for processing. They encode in order, top to bottom. This can be a preset template if needed often, or created when needed.

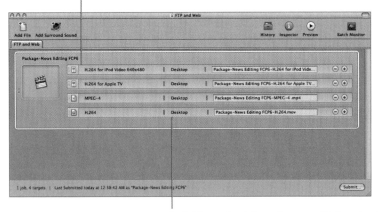

Each file can be delivered to a specific location. In this case all four encodes are going to the desktop.

You can add as many deliverables as you want for this specific story. Of course, if you are in a hurry for the primary deliverable, get that one done first, and then add the remaining jobs.

Podcasts can also have chapters assigned to them. This allows you to click the Right and Left Arrow keys when viewing the podcast on a computer, which snaps to the next and previous chapter. On an iPod, you can skip forward and back with the normal navigation buttons.

Adding markers in the Final Cut Pro Timeline is very easy. Navigate in FCP to the desired location in the Canvas/Timeline and press M once to make the marker, and a second time to add information or a name. If you only want to provide the viewer with chaptering, add more markers and do the export. Then click the Chapter Markers button to make these markers arrive in Compressor as chapters.

If you want to add rich content to the markers (such as links to web pages), you can add links within Final Cut Pro. This information will be available for viewing only on a computer and not on the iPod.

You can also add markers and link to rich content from within Compressor. In the Preview window, move the playhead to the desired location and press M.

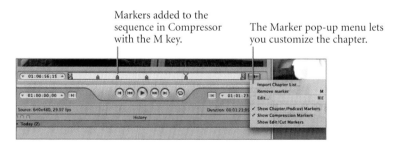

Markers added to the sequence in Compressor with the M key.

The Marker pop-up menu lets you customize the chapter.

If you select Edit, a dialog opens to determine what you want to add to that marker/chapter.

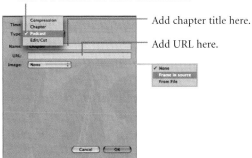

Select Podcast from the Type menu to add a URL to link to a website for more information.

Add chapter title here.

Add URL here.

This frame will be associated with the chapter marker in the QuickTime and iTunes playback control bars.

You can also assign a specific frame to the chapter marker. This shows up in iTunes or QuickTime when you are choosing the chapter from the menu. You can select from the frame where the marker is located or from another file.

Preparing Content for Apple TV

Apple TV has the ability to link directly to one computer in a home or office and display the content on an attached widescreen television or projector.

The settings in iTunes are much like syncing an iPod, in that you can tell iTunes what content you want to deliver to the device. In the case of Apple TV, you can sync all or part of your iTunes music library, all or part of your photo albums, and any or all podcasts, movies, or programs you may have downloaded with iTunes or from individual website subscriptions, provided the movies have been encoded in Apple TV format.

Podcasts are available from the iTunes Music Store specifically for Apple TV, in both SD and HD. You can also put Apple TV podcasts on your website. If you use the Apple TV settings for encoding your story and deliver it to the location for the website, subscribers to your Apple TV podcast can have your content downloaded and then transferred to their Apple TV when new material is available.

If the subscriber has the Apple TV settings configured properly in iTunes, these Apple TV podcasts are copied from the "host" computer to Apple TV whenever iTunes is launched and the subscribed podcast is downloaded.

This can provide an entirely new way of delivery of HD content for both TV stations as well as newspapers and web services, and it just happens to be built in to the Apple workflow.

Proceed with all the steps above for preparing a podcast and adding chapters.

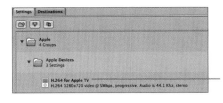

Drag the H.264 for Apple TV setting to the Batch window.

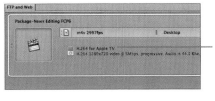

Drag the icon to the blue area of the Batch window. This creates a second "job" for the existing sequence.

Confirm the destination and submit the "job."

If your website is set up properly, viewers can subscribe to your Apple TV podcast and can have that content delivered to their Apple TV for viewing on their widescreen television whenever they choose to watch it. The Apple Remote has the ability to pause, back up, and skip forward, as well as make playback selections.

You will need to know at least one of these methods in order to get your story delivered. As you learn more about each, you may be able to persuade your operation to explore implementing some of these methods to attract viewers.

10
Customization Overview

Once Final Cut Pro's standard interface and default settings are familiar, customizing them to streamline frequent tasks can speed up the editing process considerably. As seen in the previous chapters, many FCP functions can be accomplished multiple ways, including menu commands, keyboard shortcuts, and cursor tools. FCP also allows creation of custom shortcuts.

Creating Keyboard Shortcuts

From the Tools menu you can choose a number of settings to customize.

The "coffee bean" button bar atop the Viewer

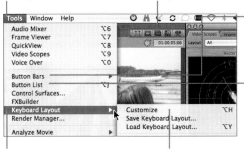

Button bars (or "coffee beans") sit at the top right of the Browser, Viewer, Canvas, and Timeline.

Choosing Button List gives you a searchable list of all possible buttons and functions that you can change or assign to other keys.

The Keyboard Layout option shows you the existing layout as well as the button list. The Button List option opens a larger window with the same info.

To open the Keyboard Layout window, choose Tools > Keyboard Layout > Customize, or press Option-H.

Use the Keyboard Layout modifier key tabs to assign multiple-keystroke shortcuts. Dragging a command shortcut to the F5 key on the Shift-Command tab, for example, means that command will be executed when you press Shift-Command-F5.

As you type in the search field, the results narrow to display only command names that include the sequence of letters you've entered.

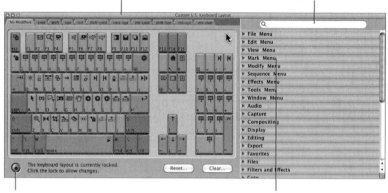

The keyboard might be locked. Click here to unlock and accept changes.

You can map any menu command to a keyboard shortcut. Most commands are already mapped to some mix of character and modifier keys.

1   When you enter a search term in this field, the results list shows every command that contains the typed letter sequence anywhere in its name. Type *wa* to search for *waveform*, and command names that contain *Forward* are listed along with the desired command. You can add a *v* to make the search term *wav* and isolate just the command name that contains *Waveform*, or just select from the longer list of results. Either way, click the Toggle Waveform Display entry in the search results list.

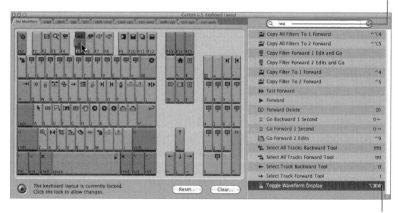

2   If Final Cut Pro already has a keyboard shortcut assigned to a listed command, you'll see the keystroke combination next to the command's name. To assign a new keyboard shortcut, drag the highlighted command name to a key on the appropriate Keyboard Layout tab. If you highlight the command in the search results, you can also press the desired target key to map that function. All associated shortcuts, including the new one, will appear next to the command name.

Creating Keyboard Shortcuts 175

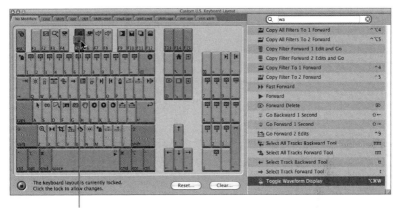

3 When you drag a command name to the Keyboard Layout map, it changes into a Key Cap icon. Drag the command name to the key you want to assign to the shortcut (in this case, it's F5).

To save a keyboard layout on your system do the following:

Use the Save Keyboard Layout command to save your custom shortcut configuration as a file for backup purposes or for transferring the settings to another Final Cut Pro workstation.

Use the Load Keyboard Layout command to open and apply a custom shortcut configuration you've saved or migrated from another workstation.

If your intention is to bring your layout setting to another system, navigate from this menu to another storage device such as a thumb drive.

If you save your custom keyboard in the Keyboard Layouts folder, it will show up in the Tools > Keyboard Layout menu. This way, you won't have to search for it or remember what you named it. If you are on an Xsan shared storage environment, you could put these into a shared folder for easy access at any of the connected FCP edit systems.

From the same menu for saving a keyboard layout, choose Load Keyboard Setting.

The Path view has been selected to show you where the custom settings files sit within the various folders

If you named your setting something specific and memorable, type it in the search field to find it.

Choose Tools > Keyboard Layouts > Load, to display this window. Navigate to a stored custom keyboard layout file to select and load it.

Configuring Button Bars

Using keyboard shortcuts is one way to quickly access certain features. You can also use the mouse to click a button that would typically require multiple keystrokes or deeper menu selections.

The "coffee bean" that appears in the upper-right corner of the FCP Browser, Viewer, Canvas, and Timeline is a pair of end brackets for an expandable button bar, to which you can add custom buttons that provide easy access to frequently used commands. In the Timeline, linked selection and snapping are turned on by default.

Configuring Button Bars 177

Use the Keyboard Layout window to search for and access commands, just as you did in the preceding section, "Creating Keyboard Shortcuts."

As you'd do to create a keyboard shortcut, drag a command name from the Keyboard Layout search-results list to the button bar. You'll see the command name turn into a button, which you can position where you like, then drop into place.

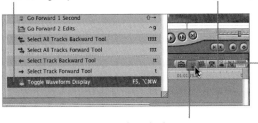

The Browser, Viewer, and Canvas all have their own button bars, which you can customize, and you can populate the same buttons in all button bars.

To rearrange, drag the button where you want it to go.

To remove a shortcut button, drag it out of the bar. It appears to vaporize.

Consider adding these often-used buttons to the Browser button bar:

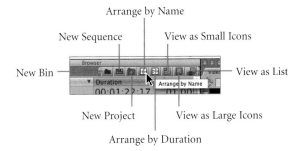

- New Bin
- New Sequence
- New Project
- Arrange by Name
- Arrange by Duration
- View as Small Icons
- View as Large Icons
- View as List

These are useful additions to the Canvas button bar:

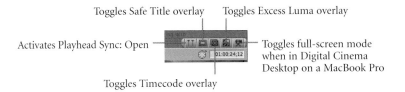

- Activates Playhead Sync: Open
- Toggles Safe Title overlay
- Toggles Timecode overlay
- Toggles Excess Luma overlay
- Toggles full-screen mode when in Digital Cinema Desktop on a MacBook Pro

Here are some handy shortcuts in the Timeline button bar:

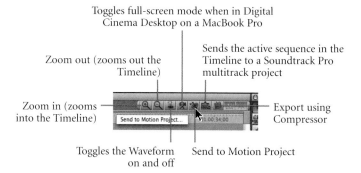

There are no limitations on which buttons go where. As new tasks become routine, remember to consider creating shortcuts for them.

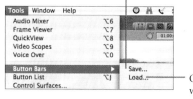

Choose Tools > Button Bars > Save to back up all custom button bars in a single step, and to create a file for use in reproducing a button bar setup on a copy of FCP installed on another Mac. This command saves all the button bars for all four windows, no matter which window is active when you apply it.

Choose Tools > Button Bars > Load when you want to use a custom button configuration file you've saved.

Name your button bars file to make it easy to find. If you save your custom button settings in the Button Bars folder, it will show up in the Tools > Button Bars menu. That way you don't have to search for it or remember the name. If working on an Xsan shared storage system, create a shared folder in which to deposit settings. That makes custom settings available from any Mac on the network.

Choose Tools > Button Bars > Load, to display this window. Navigate to a stored custom button bars file to load it.

Customizing the Window Layout

Certain parts of the editing process may require different sized windows. Final Cut Pro can store custom window settings for easy reuse.

The most common window arrangement, referred to as Standard, is displayed in the Overview section of Chapter 1. View other preset window configurations by choosing Window > Arrange.

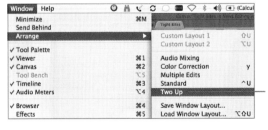

Choose Window > Arrange > Two Up to expand the Viewer and Canvas and scale down the Timeline. Assign a keyboard command or button if this is a setting you use frequently.

180 Customization Overview

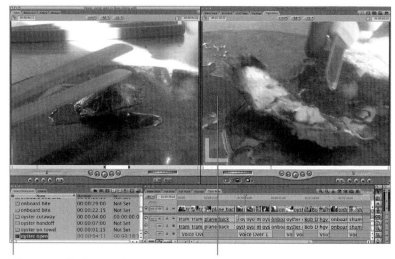

The two-up display puts the Browser next to the reduced Timeline and enlarges the Viewer and Canvas.

The Viewer and Canvas are enlarged in the two-up display.

You can bring back the Standard window layout by pressing Option-U.

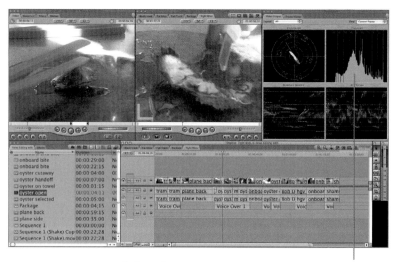

With the Color Correction preset (keyboard shortcut Y), the Waveform Monitor, Vectorscope, and RGB Parade windows appear.

Customizing the Window Layout 181

You can choose from this menu whether you want the mixer to connect to the Viewer or Canvas/Timeline. It is best to leave the Auto option so that the mixer switches as you toggle between windows.

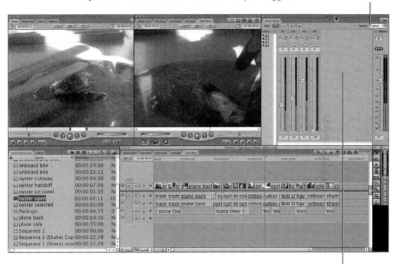

The Audio Mixer

If none of the preset window configurations suffices, adjust windows individually by hovering the mouse over a border and dragging a Resize pointer. Resizing any FCP window causes the other windows to scale as well, to compensate for the adjustment.

Drag left to right on a vertical window border to adjust window width.

Drag in any direction and the windows all resize to compensate.

Saving a Final Cut Pro custom window configuration is very similar to saving button bars and keyboard layouts.

To restore your keyboard layout, choose Window > Arrange > Load Window Layout and navigate to your saved window layout file.

As you progress with Final Cut Pro, you will find that having custom settings for the functions you use frequently will make them readily accessible and therefore second nature.

Index

Numbers
0 db cross fade, effect of, 85
+3 dB cross fade
  effect of, 85
  as standard default, 86
3G service with cell phones,
  use of, 158
23 GB XDCAM discs,
  recording time for, 21
720p30 broadcast format,
  explanation of, 15
720p60 broadcast format,
  explanation of, 15
802.11 wireless networking,
  use of, 159–160
1080i HD video, shooting
  with Sony XDCAM HD
  cameras, 21
1080ip60 broadcast format,
  explanation of, 15–16
1394, setting deck input to, 149

A
air, playing video to, 147
Apple Devices folder, setting
  iPod deliveries in, 168
Apple TV, preparing content
  for, 170–172
audio. *See also* NATSOT
  (natural sound on tape)
  applying Low Frequency
  adjustment to, 139
  targeting to tracks, 9
audio cross fades
  applying in Timeline, 86–87
  changing default for, 86
  dragging and dropping, 86
  features of, 85–86
  positioning between clips, 87
audio effects, viewing, 4
audio equalization, achieving,
  138–140
audio keyframes
  inserting and deleting, 83
  removing, 81
Audio Level meters,
  description of, 2
audio level overlay, putting
  keyframes on, 106

audio levels
  adjusting, 79–80
  adjusting in Viewer, 83
  lowering at ends of
    sequences, 111
  normalizing, 79
  ramping in Timeline, 80–82
  ramping within clips in
    Viewer, 84
audio meters, using, 10
audio render bar, availability
  of, 144
audio tracks
  changing for packages, 94
  clearing from packages,
    119–120
  looping from In to Out
    point, 139
audio waveforms
  identifying takes with, 96
  making visible in Timeline, 96

B
backtiming sequences,
  109–110
batch capture, using with
  videotape, 53
Batch Monitor in Compressor,
  opening, 161, 165
BGAN (broadband global area
  network), use of, 159
bins
  creating from File menu, 6
  displaying contents of, 4
  organizing video elements
    into, 3–4
  viewing contents of, 4–6
bites. *See* sound bites
blade keyboard shortcut,
  splitting shots with, 10
blurring subjects, 128–132
Brightness filter, adding to
  effects, 133–134
broadband story delivery
  methods
  3G service with cell
    phones, 158
  BGAN (broadband global
    area network), 159
  cable modem, 157–158

DSL (digital subscriber
  line), 157
EVDO (Evolution Data
  Optimized) cards, 158
overview of, 155–157
UMTS (Universal Mobile
  Telecommunications
  System) cards, 158
wi-fi connections, 159–160
B-roll
  adding, 103–109
  inserting for closing clip, 110
  relationship to packages, 91
B-Roll bin, opening as
  separate window, 4
B-roll clip, extending start
  of, 113
B-roll sound level, adjusting
  for natural sound pops,
  118–119
Browser
  changing options for, 5
  displaying projects in, 4
  Effects tab of, 4
  features of, 3–4
  in List view, 5
  toggling between List and
    Frame views, 3
  viewing contents of, 5
Browser button bar, adding
  buttons to, 177
Browser window, contents of, 2
button bars
  backing up customizations
    of, 178
  configuring, 176–179
  dragging command buttons
    onto, 3
  identifying, 173, 176
  loading, 179
  naming files for, 178
Button List, choosing, 173

C
cable modems, use of, 157–158
Cache button in XDCAM
  Transfer, options related
  to, 26
Canvas
  connecting mixer to, 181

Index **183**

enlarging in two-up display, 180
features of, 7–8
splitting clips in, 8
toggling between Viewer, 9
versus Viewer, 8
Canvas button bar, adding buttons to, 177
Canvas sources, locking Viewer sources to, 6
Canvas window, description of, 2
capture choices, availability in Log and Capture window, 50
Capture Now button, keyboard shortcut for, 54
clip length, identifying relative to audio levels, 83
clip mixes, auditioning, 81
Clip Overlays control, adjusting audio levels with, 80
clip position, identifying for audio levels, 83
clip previews, speeding through, 7
clips. *See also* shots
accessing, 7
adding seconds and frames to start of, 113
adjusting audio levels of, 79–80
applying speed adjustments to, 123–126
assigning default target locations for, 5
attaching comments for storyboard edits, 69
cutting with Razor Blade tool, 81
determining location in, 7
dragging and dropping in Timeline, 70–72
extending beginnings and endings of, 28
extending over narration, 106
importing on Ikegami Editcam, 44
marking, 7
marking for B-roll, 103
matching to Timeline's settings, 13
naming for videotape capture, 52
navigating, 7
navigating frame by frame, 7
opening in Viewer, 135
organizing with Browser, 3
playing and stopping in XDCAM Transfer, 35

positioning audio cross fades between, 87
ramping levels in within Viewer, 84
rearranging in Timeline, 69–72
removing from packages, 119–120
resizing, 134–135
selecting and importing in XDCAM Transfer, 30
selecting and using from Browser, 3
shifting with slide edits, 108–109
sorting by timecode values in XDCAM Transfer, 31
splitting in Canvas, 8
tailing, 75–76
topping, 74
transferring on P2 cameras, 41
viewing as frames for storyboard edits, 67
viewing in XDCAM Transfer proxy-selection window, 30
zooming out, 83
clips from videotape, using Viewer with, 72
coffee bean. *See* button bars
Color Corrector, using, 128, 180
columns
clustering, 5
sorting in XDCAM Transfer software, 32
commands, searching for and accessing, 177
comments, attaching to clips for storyboard edits, 69
compressing stories, 160–166
compression and transfer speeds for stories, examples of, 166–167
compression groups and formats, displaying, 16
compression rates, selecting for videotape capture, 51
compression settings, specifying in Compressor, 161
Compressor
adding markers in, 169
linking to rich content from, 169
using for FTP, 162
Compressor droplets, use of, 160
cross dissolve, description of, 89
cross fades. *See* audio cross fades

D

Device Control settings, specifying for videotape capture, 51
Direct To Edit FireStore recorders, features of, 20
dissolves
adding, 88–90
moving, 90
removing, 90
droplets, compressing stories with, 160
DSL (digital subscriber line) data transfers allowed by, 156
sample transfer times for, 167
use of, 157
DV25 format, explanation of, 18
DVCAM format
explanation of, 18
size of data file yielded by, 155
DVCPRO 25 format
sample compression speed for, 167
size of data file yielded by, 155
DVCPRO HD format
description of, 14
playing back edited stories from, 149, 154
storage consumed by, 155
DVCPRO SD format, explanation of, 17

E

Easy Setup, opening to export to playout server, 12
edit options, displaying for storyboard edits, 68
Edit to Tape option, accessing, 148
Editcam FieldPak
ejecting, 43
mounting for Editcam content, 43
edits
extending into black, 111
overwriting, 111
undoing, 71
effects. *See also* real-time playback of effects
identifying defaults for, 4
playing in real time, 136
rendering directly to shared storage, 146
Encoder information, displaying in Compressor, 164

Index

encoders, use with satellite trucks, 149
EQ filters, using, 138–140
EVDO (Evolution Data Optimized) cards
  sample transfer times for, 167
  use of, 158

F

Fat Bites step of laying out packages, explanation of, 92
FCS (Final Cut Server). *See* Final Cut Server (FCS)
FieldPak
  ejecting, 43
  mounting for Editcam content, 43
File menu, creating bins and sequences from, 6
filters, blurring subjects with, 129–130
Final Cut Pro, windows in, 2
Final Cut Server (FCS)
  accessing, 63
  Assets window in, 58
  creating projects in, 58–59
  features of, 57
  managing editor's workflow with, 62–64
  managing producer's and reporter's workflow with, 58–62
  playing back proxy views in, 60
  shot selection tool in, 60
FireStore media-capture device, features of, 45–46
FireStore recorders, features of, 20
FireWire NTSC Device Control, setting for videotape capture, 51
fit to fill edits, performing, 121–123
Flip4Mac Editcam
  delivering material to bins from, 45
  features of, 43–44
formats
  HD (high definition), 14–17
  SD (standard definition), 17–18
formatting, effect on XDCAM disks, 22
Frame and List views, toggling Browser between, 3
frame rates
  choosing for videotape capture, 51
  choosing in Easy Setup window, 17
frame-by-frame navigation, using, 7
frames
  adding to sound bites, 109–110
  arranging for storyboard edits, 67
  assigning to chapter markers for podcasts, 170
  navigating in Log and Capture window, 49
  placing marks in, 9
  selecting for sliver clips, 117
  shifting with slip edits, 107–108
  size of full frame rate, 164
  viewing clips as for storyboard edits, 67
FTP
  sample transfer times for, 167
  using Compressor for, 162, 164
FTP servers, setting up for broadband story delivery, 156
FTP utilities, examples of, 156
full frame rate, size of, 164
Full Track step of laying out packages, explanation of, 92, 95–97

G

gaps, filling with shots, 107
Gaussian Blur filter, using, 129
generator files, accessing, 7
Genlock source option, using with Matrox MXO converter box, 153

H

HD (high definition) format, overview of, 14–17
HD compression settings, mixing, 22
HD frames, filling 4x3 space in, 12
HD standards, delivery formats supported by, 15
HD video, playing back edited stories from, 149
HDCAM format, explanation of, 14
HDV format, explanation of, 14
HDV Timeline, playing to tape in Edit to Tape mode, 148
highlighting subjects, 133–134
HQ (high quality) XDCAM setting, FCP setting for, 21

I

Ikegami Editcam
  editing content on, 43
  features of, 43–45
image stabilization, achieving with SmoothCam, 137–138
images, resizing with Scale slider, 135
IMX format, explanation of, 17
In points
  finding, 7
  marking, 7, 9
  marking for storyboard edits, 69
ingest stations, use of, 47
Inmarsat, satellites administered by, 159
Insert Edit button in Canvas, effect of, 8
insert edit mode, performing shuffle edit in, 70–71
iPod 320 × 240, choosing as default compression setting, 161–162
iPod deliveries, adding settings for, 168

J

jog control, identifying, 7, 49

K

Kb (kilobit), measurement of, 155
keyboard layouts
  restoring, 181
  saving on system, 175
keyboard shortcuts
  audio cross fade default, 86
  Capture Now, 54
  Clip Overlays control for adjusting audio levels, 80
  Color Correction preset, 180
  creating, 173–176
  creating bins, 6
  creating sequences, 6
  default effects, 4
  dissolves, 88
  Easy Setup, 12
  Final Cut Server (FCS), 63
  finding In and Out points with, 7
  keyframes, 7
  linked selection, 69, 105

Index **185**

Log and Capture window, 49
mapping menu commands to, 174
Overwrite button, 104
overwriting edits, 111
Pen tool, 10, 84
for playing and stopping clips in XDCAM Transfer, 35
Razor Blade tool, 81
Razor Blade tool used to clean up narration tracks, 97
ripple edit, 9
Ripple tool, 105
roll edit, 9
Selection tool, 71
Sequence Settings, 13
Show Audio Waveforms, 95
slide edits, 10
slip edits, 10
Slip tool, 108
snapping, 69
Standard window, 180
subclips, 98
tailing clips, 75–76
Tool palette, 9–10
undoing actions, 70
undoing edits, 71
Voice Over tool, 93
zooming out clips, 83
keyframes
 adding, 7
 adding relative to backtiming, 110
 controlling amount of blur with, 131
 dragging relative to audio levels, 84
 putting on audio level overlay, 106
kilobit (Kb), measurement of, 155

L

linked selection
 turning on, 69
 using with pad at top of stories, 105
List and Frame views, toggling Browser between, 3
List view, displaying Browser in, 5
live to air, playing sequences from, 147
Log and Capture window
 features of, 49–50
 opening for videotape, 49
Logging bin
 contents of, 5
 identifying, 4

Logging tab, using with videotape capture, 52
logging videotape prior to capture, 48
Long GoP (Long Group of Pictures), explanation of, 14

M

Macs, connecting XDCAM devices to, 28
markers
 adding, 7
 adding in Timeline, 169
 adding rich content to, 169
marks
 displaying in List view, 5
 placing in frames, 9
Mask Shape filter, using, 131, 133
Matrox MXO converter box, setting up, 150–155
Mb (megabits), measurement of, 155
MB (megabyte), measurement of, 155
media, specifying delivery for XDCAM Transfer software, 25
media management capabilities, adding with FCS (Final Cut Server), 57
Media Map column on P2 cameras, explanation of, 40
megabits (Mb), measurement of, 155
megabyte (MB), measurement of, 155
menu commands, mapping to keyboard shortcuts, 174
metadata
 benefits of, 22
 specifying delivery for XDCAM Transfer software, 25
mini DV tape, using, 56
mixer, connecting to Viewer or Canvas/Timeline, 181
mono sounds, moving between speakers, 84
motion effects
 performing fit to fill edits, 121–123
 time remapping, 126
MPEG-4 compression setting, choosing in Compressor, 164

N

narration
 adding video to, 103–109
 backtiming and extending sound accent for, 117–119
 recording for packages, 93–95
Narration bin, determining contents of, 4
narration levels, checking, 10
narration tracks
 cleaning up, 96–97
 opening up for natural sound pops, 115–119
NATSOT (natural sound on tape). *See also* audio
 opening for, 112–115
 using timecodes for, 55
No Recording Range Set message, dealing with, 94
normalization, applying to audio levels, 79
NTSC Device Control, setting for videotape capture, 51

O

Open Format Timeline, features of, 11–12
Open Timeline auto-detect feature, rejecting, 16
Option key, turning on insert edit mode with, 70
Out point marker, moving when topping clips, 74
Out points
 finding, 7
 marking, 7, 9
 marking for storyboard edits, 69
 moving to playhead, 104
overlays, displaying relative to backtiming, 109
Overwrite button, keyboard shortcut for, 104
overwrite edits, impact of, 8, 71, 111

P

P2 cameras
 changing clip transfer order on, 41
 features of, 36–38
 importing clip elements on, 42
 Log and Transfer window in, 39
 previewing shots on, 39–42
 selecting clips for transfer on, 40
 transferring clips on, 41

186 Index

P2 cards
　availability of, 37
　ejecting, 40, 43
　importing clips on, 41
P2 clip thumbnails, resizing, 41
P2 content, importing into Final Cut Pro, 38
P2 viewer, anatomy of, 42–43
packages
　completing, 119–120
　components of, 91
　laying out, 92–93
　placing first clips in, 106–107
　recording narration for, 93–95
　relationship to VO (voiceover) and SOT (sound on tape), 65
pad, adding at top of stories, 105–106
padding ends of sequences, 110–112
pan keyframes, inserting and deleting, 83
pan settings
　adjusting for speakers, 83
　adjusting for tracks relative to audio levels, 84
Pan slider, using with audio-level adjustments, 83
Panasonic P2 cameras. See P2 cameras
pans, saving relative to audio levels, 83
PC Network mode, releasing decks and cameras from, 25
PDZK-P1 download link, locating, 24
Pen tool
　keyboard shortcut for, 10
　using to ramp clip audio levels in Timeline, 80–81
　using with alternate first shot method, 106
　using with natural sound pops, 118–119
　using with ramped audio levels within clips in Viewer, 84
PictureReady software, using with videotape, 56
playback. See real-time playback of effects
playhead
　determining clip and sequence locations with, 7
　dragging, 7

extending, 10
moving Out points to, 104
playout server, exporting to, 12
podcasts
　availability of, 171
　delivering stories to, 168–170
producer's workflow, managing with FCS (Final Cut Server), 58–62
proxy files
　specifying delivery for XDCAM Transfer software, 26
　using with Sony XDCAM and XDCAM HD, 20
proxy frame size, adjusting in SDCAM Transfer software, 26
proxy selection window, appearance in XDCAM Transfer, 30
proxy transfers, allocating space in XDCAM Transfer, 26
proxy views, playing back in FCS (Final Cut Server), 60

R

radio cut
　creating, 102–103
　relationship to packages, 93
Razor Blade tool
　cleaning up narration tracks with, 97
　cutting clips with, 81–82
real-time playback of effects. See also effects
　complexity of, 145
　considerations related to, 143–146
reel name and number, identifying for videotape, 53
render bars, availability of, 144
rendering, considering in real-time playback of effects, 143
reporter's workflow, managing with FCS (Final Cut Server), 58–62
rewrapping clips, occurrence of, 19
ripple edit
　explanation of, 102
　keyboard shortcut for, 9
Ripple tool
　using with opening for NATSOT, 113
　using with pad at top of stories, 105

roll edit, keyboard shortcut for, 9
rough-cut edits, accessing in FCS (Final Cut Server), 63
rough-cut sequences, assembling in FCS (Final Cut Server), 60
RT (real time) modes, activating, 144

S

Safe RT option
　selecting, 144
　switching to, 145
satellite trucks, playing Timeline to, 149–155
Save Keyboard layout command, using, 175
Scale slider, resizing images with, 135
SD (standard definition) format
　overview of, 17
　sending to encoder, 155
SD material, enlarging, 12
SD video, editing, 149
Selection tool, keyboard shortcut for, 9
selections
　extending, 111
　rotating through, 113
sequence content, considering for real-time playback of effects, 143–146
sequences
　assigning default target locations for, 5
　backtiming, 109–110
　compressing in Timeline, 160
　creating from File menu, 6
　determining location in, 7
　displaying in Timeline tabs, 9
　displaying in Unlimited RT mode, 145
　gaps in, 71
　opening across tracks, 116
　organizing with Browser, 3
　padding ends of, 110–112
　playing back to videotape, 148
　playing in real time, 146
　playing live to air, 147
　previewing, 7
　zero-duration label for, 4
servers, editing on, 146
Set Logging Bin command, effect of, 5
shortcut buttons, removing, 177

Index **187**

shot selection tool in FCS (Final Cut Server), using, 60
shots. *See also* clips
breaking in two pieces, 10
filling gaps with, 107
previewing on P2 cameras, 39–42
selecting on XDCAM discs, 33
warming up, 128
Show Audio Waveforms option, keyboard shortcut for, 95
shuffle editing
performing, 71
performing in insert edit mode, 70
rearranging clips in Timeline by means of, 69–72
slide edit
keyboard shortcut for, 10
performing, 108–109
slip edit
keyboard shortcut for, 10
performing, 107–108
sliver clips, selecting end frames of, 117
SmoothCam, achieving image stabilization with, 137–138
snapping
active status of, 70
turning on and off, 9, 69
using with sound bites and standups, 98
software, considering for real-time playback of effects, 143
Sony XDCAM and XDCAM HD. *See* XDCAM devices
Sony XDCAM menu option, confirming, 23
SOT (sound on tape), versus VO (voiceover), 65
sound accent
backtiming and extending, 117–119
overlapping and blending with narration, 117–119
sound bite subjects, introducing, 109–110
sound bites
adding frames to, 109–110
applying EQ to, 138
dragging to Timeline, 99–101
inserting in stories, 97–99
source material type, checking after editing shots, 13

SP (standard play) XDCAM setting, FCP setting for, 21
speakers
adjusting pan values for, 83
moving mono sounds between, 84
speed adjustments, applying to clips, 123–126
Spinback 3D effect, using, 136–137
splice edit, performing, 8
standard definition (SD) format. *See* SD (standard definition) format
Standard window, bringing back, 180
standups, inserting in stories, 97–99
stories
adding pads at tops of, 105–106
cleaning up Full Track for, 95–97
compressing, 160–166
inserting sound bites and standups in, 97–99
recording narration for, 93–95
story delivery, compression and transfer speeds for, 166–167
story delivery methods
broadband story delivery, 155–160
playing Timeline to satellite truck, 149–155
video podcasts, 168–170
to websites, 167
storyboard edits, performing, 66–68
Subclip handles setting, explanation of, 28
subclips, using with sound bites and standups, 98
subjects
blurring, 128–132
highlighting, 133–134

T

tailing clips, use of, 73, 75–78, 102–103
takes, identifying with audio waveforms, 96
tape. *See* videotape
tapeless devices. *See also* Ikegami Editcam; P2 cameras; XDCAM devices
features of, 19–20
FireStore media-capture device, 45–46
Ikegami Editcam, 43–45

tapeless media, using Viewer with, 73
thumbnails, displaying in Frame view, 3
Tight Bites, trimming to, 93, 102–103
Tight Track, using with narration, 92, 96
Time Remap controls, using, 123–124
timecode
adding in XDCAM Transfer software, 27
determining position of, 7
displaying in List view, 5
entering for videotape capture, 56
selecting, 6
sorting clips by, 31
Timeline
adding markers in, 169
applying cross fades in, 86–87
compressing sequences in, 160
connecting mixer to, 181
dragging bites to, 99–101
features of, 8–9
making audio waveforms visible in, 96
matching to source material, 12
playing, 7
playing to satellite trucks, 149–155
ramping clip audio levels in, 80–82
rearranging clips in, 69–72
toggling between Viewer, 9
topping and tailing clips with, 76–78
visual representation of contents of, 7–8
Timeline button bar, shortcuts in, 178
Timeline settings
matching clips to, 13
retaining, 13
Timeline window, description of, 2
Tool palette
description of, 2
features of, 9–10
topping clips, 74, 76–78, 102–103
explanation of, 73
versus trim mode, 78
tracks
displaying in List view, 5
locking, 9
trailing clips, versus trim mode, 78

transitions
  adjusting with roll edits, 9
  applying for storyboard
    edits, 68
  cycling through, 10
  preventing gaps between, 81
  two-up window display,
    Viewer and Canvas
    enlarged in, 180

U
U-Matic, capabilities of, 17
UMID (Unique Media ID)
  option in XDCAM
  Transfer software,
  explanation of, 27
UMTS (Universal Mobile
  Telecommunications
  System) cards, use of, 158
undoing actions, 70–71
Unlimited RT mode
  displaying sequences in, 145
  effect of, 146
Use pop-up for formats,
  options in, 17

V
Variable Speed option,
  applying, 126
Vectorscope
  toggling visibility of, 52
  using in Log and Capture
    window, 50
video, adding to narration,
  103–109
video capture, avoiding pre-
  roll during, 53
video effects, viewing, 4
video elements, organization
  into bins, 3–4
video podcasts, delivering
  stories to, 168–170
video render bar, availability
  of, 144
video transitions, default
  effects for, 4
videotape
  avoiding shuttling mini DV
    tape, 56
  capturing from, 54–56
  formatting at 640 × 480
    pixels in Compressor,
    162
  identifying reel name and
    number for, 53
  inserting natural sound on,
    112–115
  locking, 105

logging prior to capture, 48
playback to, 148
playing to air, 147
preparing to capture from,
  49–54
using batch capture with, 53
using PictureReady software
  with, 56
videotape content, capturing
  to hard disk in real
  time, 47
videotape workflow,
  accelerating, 48
Viewer
  adjusting audio levels in, 83
  versus Canvas, 8
  connecting mixer to, 181
  enlarging in two-up display,
    180
  features of, 6–7
  opening clips in, 135
  ramping audio levels within
    clips in, 84
  toggling between Canvas/
    Timeline, 9
  topping and tailing clips
    with, 76–78
  using with clips from
    videotape, 72
  using with tapeless media, 73
Viewer sources, locking to
  Canvas sources, 6
Viewer window, description
  of, 2
Voice Over tool, keyboard
  shortcut for, 93
VOs (voiceovers)
  prepping, 72
  versus SOT (sound on
    tape), 65

W
Waveform Monitor
  toggling visibility of, 52
  using in Log and Capture
    window, 50
waveforms, activating for Full
  Tracks, 95
websites
  delivering stories to, 167
  for XDCAM Transfer
    software, 23
white balance, fixing, 127–128
wi-fi connections, use of,
  159–160
window layout, customizing,
  179–181
windows, resizing, 181
wireframes, selecting, 6

X
XDCAM devices
  connecting to Macs, 28
  setting up for first time,
    24–25
XDCAM discs
  ejecting, 36
  mounting, 32–34
  renaming, 33
XDCAM disks
  effect of formatting on, 22
  renaming, 22
XDCAM format, features of,
  20–23
XDCAM HD format
  capturing, 14
  features of, 20–23
  initial setup for, 23–24
XDCAM HD Timeline,
  playing to tape in Edit to
  Tape mode, 148
XDCAM SD, initial setup for,
  23–24
XDCAM Transfer clip proxy
  icon, anatomy of, 32
XDCAM Transfer software
  adding timecode numbers
    in, 27
  components of, 30–31
  depositing high-res material
    in, 27
  downloading, 23
  Import and Import All
    options in, 29
  opening from within Final
    Cut Pro, 29
  playing and stopping clips
    in, 35
  running as standalone
    application, 29
  selecting clips in, 29
  selecting portions of clips
    in, 29
  setting up, 25–28
  sorting clips by timecode
    values in, 31
  sorting columns in, 32
  UMID option in, 27
  using within Final Cut Pro,
    34–35
  viewing clips in proxy-
    selection window of, 31
XDCAM-native HD video,
  playing, 150
Xsan servers, editing on, 146

Z
zooming out clips, 83
zooms, animating, 135